Advances in Electrocatalysis

Edited by Viorica Parvulescu and Maria Marcu

Published in London, United Kingdom

Advances in Electrocatalysis
http://dx.doi.org/10.5772/intechopen.1008580
Edited by Viorica Parvulescu and Maria Marcu

Contributors
A.A. Musa, Abayomi Omoogun, Abubakar Tahir Isa, Aliyu Tahir Isah, Carlos R. Cabrera, Gabriela Petcu, Hauwa Mohammed Mari, Ilias Stamatelos, Kurt L. Macri, Magdalena Wajrak, Min Wei, Ruma Rani Paul, Subhajit Bhunia, Suzatra Chatterjee, Viorica Parvulescu, Worawee Saei, Xia Sheng

First published in London, United Kingdom, 2026 by IntechOpen
IntechOpen is the global imprint of INTECHOPEN LIMITED, registered in England and Wales, registration number: 11086078, 167-169 Great Portland Street, London, W1W 5PF, United Kingdom

For EU product safety concerns: IN TECH d.o.o., Prolaz Marije Krucifikse Kozulić 3, 51000 Rijeka, Croatia, info@intechopen.com or visit our website at intechopen.com.

British Library Cataloguing-in-Publication Data
A catalogue record for this book is available from the British Library

Advances in Electrocatalysis
Edited by Viorica Parvulescu and Maria Marcu
p. cm.
Print ISBN 978-1-83635-451-2
Online ISBN 978-1-83635-450-5
eBook (PDF) ISBN 978-1-83635-452-9

Meet the editors

Viorica Parvulescu is a Senior Romanian Academy Researcher in the Chemistry of Surface and Catalysis Laboratory at the Institute of Physical Chemistry-Ilie Murgulescu. After earning a Ph.D. in Chemistry from the Doctoral School of the Romanian Academy, she held a postdoctoral fellowship at the University of Namur in Belgium. She was a ten-year Associate Professor at the University of Bucharest. She developed expertise in preparing catalysts. Her research interests are centered on structure-property relationships, the development of new nanocomposite materials (including transitional metals, enzymes, organometallic complexes, and metallic nanoparticles supported on polymers and nanostructured oxides), and organic-inorganic hybrid materials with catalytic, photocatalytic, and electrophotocatalytic properties. She is an author and co-author of over 200 publications and seven chapters in six books.

Maria Marcu is a Senior Romanian Academy Researcher in the Electrochemistry and Corrosion Laboratory at the Institute of Physical Chemistry - Ilie Murgulescu. She received her Ph.D. degree in Chemistry in 1997 from the Doctoral School of the Romanian Academy. From 2002 to 2020, she was an Associate Professor at the National University of Science and Technology Politehnica from Bucharest. Her research interests focus on the synthesis of hybrid and composite materials, their electrochemical characterization, and applications in electrocatalysis and energy storage. She also investigates the corrosion behaviour of metallic biomaterials in biofluids. She has authored or co-authored over 70 publications, three chapters, and two books.

Contents

Preface

Electrocatalysis represents one of the most dynamic and promising fields in contemporary chemical and energy research. The development of efficient and selective catalysts is crucial for transitioning to a sustainable energy system based on renewable resources. Fundamental processes, such as water splitting, carbon dioxide reduction, and oxygen shift reactions, depend directly on the performance of electrocatalytic materials. In a global context where the transition to clean, sustainable energy sources has become imperative, understanding the fundamental mechanisms and optimizing electrocatalytic materials is more relevant than ever. However, significant challenges remain that hinder the widespread applicability of electrocatalysis. These include limited yields, the difficulty of understanding mechanisms at the atomic level, and the high cost of noble metals, often used in electrode fabrication. In recent years, however, advances in characterization techniques, nanoscale material design, and computational modeling have accelerated discoveries in the field.

Electrocatalysis is the field of electrochemistry in which the use of a catalyst significantly modifies electrochemical redox processes, increasing speed and/or selectivity. Electrocatalysis plays a crucial role in advancing clean energy and chemical production, and is a rising star in the field of catalytic technologies. Over the last decade, the number of publications on electrocatalysis has increased, reflecting a growing scientific interest that is linked to the impetus to develop technologies based on renewable energy sources, primarily photo-, electro-, and plasma-catalysis. This research primarily focuses on water splitting and its related aspects, including the oxygen evolution reaction (OER), the oxygen reduction reaction (ORR), and the hydrogen evolution reaction (HER). Additionally, CO_2 reduction (RR) and nitrogen reduction (NRR) are significantly represented in current studies on electrocatalysis.

Interest in electrocatalysis has grown rapidly in recent years, driven by the urgent need for sustainable energy and chemical solutions to combat climate change and reduce dependence on fossil fuels. Research is now focused on understanding the complex electrochemical interface, extending beyond the catalyst material to include interactions among the electrode, electrolyte, and external fields. Key areas of investigation include the development of new materials and the enhancement of the selectivity and stability of catalysts to facilitate the design of new application systems. Significant interest is given to the synthesis of metal nanoparticles (NPs) that act as active sites for recording reactions, while supports ensure electrical conductivity, a narrow distribution of NPs, and stability in (electro)chemical environments. Significant progress has also been made in the study of elementary processes at the molecular level, including the mechanisms and diversity of reactions, as well as their applications. Applications of particular interest for electrocatalysis include clean energy storage and generation, as well as the synthesis of organic and inorganic compounds.

This book, *Advances in Electrocatalysis*, aims to provide readers with a comprehensive overview of the latest discoveries and innovations in the field. The contributions gathered here encompass multiple perspectives, ranging from the mechanistic study of electrochemical processes and the design of new catalytic materials to their applications in emerging technological systems. Each chapter is authored by leading researchers who combine experimental expertise with theoretical analyses and advanced simulations, thus offering a balance between fundamental understanding and practical application. This book addresses researchers, PhD students, and engineers who are interested in recent advances and long-term perspectives in electrocatalysis.

The primary objective of this book is to present the fundamental aspects of electrocatalysis, including new materials and methods for electrode fabrication, and to highlight their role in catalytic processes of interest, such as clean energy, environmental, and electrocatalytic synthesis. The book is organized into three sections: "Catalytic Materials and Electrochemical Processes in the Development of Electrocatalysis", "New Methods of Electrode Preparation for Electrocatalytic Applications", and "Electrocatalysis for Clean Energy Applications".

The first section introduces the heterogeneous catalysts as materials used in electrocatalysis. It provides a comparative analysis of catalytic and electrochemical processes in fuel cells, highlighting the synergies, similarities, and differences between heterogeneous and electrochemical catalysis, and their implications for future technologies. The fundamentals of electrocatalysis and the electrocatalytic phases used are exemplified for reactions of great current interest, such as the electrocatalytic reduction of CO_2 and the alkaline oxygen evolution reaction.

The second section focuses on new methods for electrochemically depositing metal nanoparticles onto supports, such as Radio Frequency Magnetron Sputtering and Rotating Disk Slurry Electrodeposition. The obtained electrodes are evaluated in electrocatalytic reactions, including the oxygen reduction reaction (ORR) and the conversion of carbon dioxide (CO_2) to ethylene (CO_2RR).

The third section discusses one of the most studied reactions, water splitting, as a sustainable energy source. While it utilizes the abundant resource of the ocean to create a clean energy carrier, direct seawater splitting faces major challenges, including the corrosive effects of chloride ions, slow reaction kinetics, catalyst instability, and the need for efficient electrocatalysts to facilitate the reactions. Their effects, both now and in the future, on the strategy for developing electrocatalytic materials with high selectivity and stability are discussed.

Viorica Parvulescu
Surface Chemistry and Catalysis Laboratory,
Institute of Physical Chemistry – Ilie Murgulescu,
Romanian Academy Bucharest,
Bucharest, Romania

Maria Marcu
Electrochemistry and Corrosion Laboratory,
Institute of Physical Chemistry – Ilie Murgulescu,
Romanian Academy Bucharest,
Bucharest, Romania

Section 1

Catalytic Materials and Electrochemical Processes in the Development of Electrocatalysis

Chapter 1

Development of Electrocatalysis in Synergy with Heterogeneous Catalysis and Progress of Nanomaterials

Viorica Parvulescu and Gabriela Petcu

Abstract

The field of electrocatalysis has made significant progress in recent years, driven by its relevance to both current and emerging applications. This chapter presents the similarities and differences between heterogeneous catalysis and electrocatalysis, highlighting the synergistic action of these processes. Thus, electrocatalytic reactions are presented in comparison with heterogeneous catalytic processes and the performances obtained in the synthesis of nanomaterials. The role of nanostructured materials in improving electrocatalytic activity, stability, and selectivity is highlighted, with special emphasis on their applications as electrodes and supported catalysts. Various applications of these materials as catalysts and electrolytes in the production of hydrogen or oxygen, oxygen reduction, carbon dioxide, and nitric oxide reduction or methanol oxidation reaction are presented. Since solid electrolyte fuel cells are among the most promising solid-state electrochemical systems, this chapter presents the materials used in the fabrication of the electrodes of these fuel cells. Furthermore, for these systems, the influence of three-phase boundaries, material morphology, and surface interactions on catalytic performance is evaluated. Finally, the chapter examines the synergistic relationship between heterogeneous and electrochemical catalytic processes, highlighting their similarities and differences.

Keywords: electrocatalysis, heterogeneous catalysis, nanomaterials, anode, cathode, electrolyte, applications, fuel cells

1. Introduction

Heterogeneous catalysis and solid-state electrochemistry have traditionally been regarded as separate fields within physical chemistry, although similar conceptual approaches have often been employed to model the kinetic behavior of these processes [1]. Heterogeneous catalysis is based on surface reactions, which require the adsorption of at least one of the reactants on the catalyst surface. Electrocatalysis is a catalytic process that occurs at the surface of an electrode through a net charge transfer, being an important branch in the development of electrochemistry. An electrocatalyst facilitates a chemical reaction on the electrode surface, acting as a catalyst involved in electrochemical processes [2–4]. Therefore, electrocatalysis consists in the study of the catalytic material used at the electrode surface or as electrode in an electrochemical reaction. Heterogeneous catalysis and electrocatalysis can work synergistically together, influencing the reaction rate, efficiency, and selectivity of the processes through various mechanisms. Also, surface science, developed as a related field to heterogeneous catalysis, provides similar investigative techniques to both fields to study their fundamental aspects at the molecular level. In recent years, it has been demonstrated that, like the electrochemical rate, typical of the electrocatalytic process that occurs through a net charge transfer, the catalytic rate of catalysts in contact with a solid electrolyte also depends on the electric potential. Therefore, between heterogeneous catalysis and electrocatalysis there are many similarities [5]. Both processes occur through successive steps, such as absorption, the transformation of species adsorbed on the surface of the solid catalyst or electrocatalyst into intermediates, and reaction products that are finally desorbed. After adsorption, the reaction occurs through a series of steps, which may include the formation of active intermediates through the participation of the active centers of the catalyst or through charge transport, processes that lead to the formation of new bonds and the generation of reaction products that are desorbed. It is thus observed that the difference between heterogeneous catalysis and electrocatalysis is given by the type of process that determines the formation of intermediates and reaction products. In the first case, the adsorbed reactants are chemically activated by the catalyst through the active centers and in the second, a charge transport is carried out. Therefore, in heterogeneous catalysis, the catalyst facilitates the reaction through chemical interactions, lowering the activation energy. In electrocatalysis, an applied electrical potential drives the reaction by influencing the transfer of electrons between the reactants and the catalyst's surface. Thermodynamically, similar to a catalyst, an electrocatalyst affects the activation energy of a reaction. In heterogeneous catalysis, the activation energy is determined by the intrinsic properties of the catalyst surface and the interaction with the reactants, while in electrochemical reactions it is related to the applied potential or voltage. Because electrochemical reactions involve the transfer of electrons between chemical species, these transformations are affected by the interactions with active sites on the electrode surface, which lower the voltage needed for the reactions to occur. Stability, selectivity, and activity are, as for catalysts, three aspects that determine the evaluation of electrocatalysts. At a specific applied voltage, the activity of an electrocatalyst can be measured by calculating the resulting current density, which can then be used to determine the reaction rate through the Tafel equation [6, 7]. This equation is essential in electrochemical kinetics and electrocatalysis, as it establishes a quantitative relationship between current and applied electrochemical potential providing insight into the kinetics and mechanism of electrochemical reactions at electrode surfaces.

In addition to the similarities, electrocatalysis also has unique advantages compared to heterogeneous catalysis. It allows convenient and efficient control over the direction and rate of a reaction at ambient pressure and temperature, by adjusting the electric field of the solid-liquid interface, thereby altering the activation energy. This explains the increasing interest in recent years for applying electrocatalysis technologies in energy conversion and chemical synthesis. Looking at the comparison of heterogeneous catalysis and electrocatalysis, it is observed that in addition to the presented similarities, there are also some different characteristics of each one. Thus, the heterogeneous catalytic reaction is controlled by regulating parameters such as temperature, pressure, and concentration of reactants, while the electrocatalytic reaction by precisely controlling the electrode potential. Also, if in the case of the catalytic reaction the transfer of electrons usually takes place at the molecular level between the adsorbed reactants and the catalyst surface, in the case of electrocatalytic processes, fast and long-distance transport of electrons from the active centers in the catalyst to the electrode is necessary. In addition, high electrical conductivity is essential [5].

The electrocatalytic reaction involves multi-electron transfer reactions and adsorbed intermediates within the double layer at the electrode-electrolyte interface. Therefore, a theory of electrocatalysis must account for the effects of the double layer, electron transfer, chemisorption, and particularly their interactions [5, 8–10]. The double layer formed at the interface between a solid electrode and a liquid solution is considered the heart of electrochemistry and was initially conceptualized by Hermann von Helmholtz in 1853 [8]. In heterogeneous catalysis, the "double layer" refers to the electrical double layer that forms at the interface between the surface of a catalyst and a solution or gas. This layer is pivotal in determining the activity, stability, and selectivity of the catalyst. Compared to electrochemical processes, electrocatalytic double layers (EDL) have different properties. A key feature of these is the presence of chemisorption and chemisorbates on the electrode surface, which can carry a partial charge resulting from electronic interactions between the electrode and the adsorbent. The EDL simultaneously governs both the inherent activity and the stability of the nanostructure in supported catalysts. In 2022, Prof. Vayenas [11] analyzed the metal-support interaction (MSI) as a phenomenon closely related to Electrochemical Promotion of Catalysis (EPOC) or Non-Faradic Electrochemical Promotion Effect of Catalysis (NEMCA). It is thus indicated that ongoing research investigations are focused on two primary directions: the development and evaluation of larger and more effective multiple electropromoted systems and the electrochemical promotion of the monodispersed catalysts. Twenty years ago, Vayenas and co-workers [12] presented the progress made in both elucidating the molecular mechanism and promoting electrochemical processes in monolithic catalytic reactors. It was shown that electrochemical promotion of catalysis and metal-support interactions are phenomena that affect both chemisorption and catalytic properties.

Electrocatalysis and heterogeneous catalysis involve elementary processes and common mechanisms in the transformation of molecules. Although there are synergies and similarities in the action of the materials used, some differences have also been highlighted [12–14]. The main difference is that the support materials must to be highly conductive in electrocatalysis, whereas heterogeneous catalysts do not require this for many reactions. In this case the support materials primarily serve a physical or dispersing role. In both areas, heterostructured catalysts with multiple components are used to achieve synergistic effects and improved performance. Therefore, the synthesis of these materials requires different preparation methods, with electrocatalysts often requiring complex, mechanically designed heterostructures, while heterogeneous catalysts can be simpler and prepared under less demanding thermal conditions [15].

However, current directions in heterogeneous catalysis focus on energy conversion strategies, thus demonstrating a convergence with the directions and principles of electrocatalysis. Several shared principles and mechanistic characteristics are starting to become apparent between the two fields. Therefore, advances in both the science and applications of these two fields are based on comprehending both the synergies and the differences between them.

Studies on the clear links between electrocatalysis and heterogeneous gas-phase catalysis have been successfully complemented by the increasing efforts in recent times to perform heterogeneous catalysis in solution. These new results have linked heterogeneous catalysis to the principles and phenomena of electrocatalysis.

Advances in the development and applications of the heterogeneous catalytic and electrocatalytic processes have been achieved by combining theoretical and experimental studies [8, 10]. Currently, the progress of both experimental and theoretical methods allows us to visualize the surface reactions at the atomic scale, to understand and predict the action of heterogeneous catalysts [16]. These results also provided insights into electrochemical transformations and guided the development of high-performance electrocatalysts by increasing the number of active sites and their activity.

As it is challenging to elucidate the catalytic mechanism and identify the true active sites in electrocatalysis by experimental study alone, recently, the computational approach employing density functional theory (DFT) has become an indispensable modeling tool across numerous research areas [17]. First of all, the theoretical calculations are carried out for the design of high-performance catalysts by identifying important parameters such as the relationship between structure and performance. At the same time, experimental testing and characterization are assisted by calculations of changes in energy, electronic transitions, as well as the molecular and electronic structures during the catalysis process.

In this chapter, we will present the development of electrocatalysis highlighting the similarities with heterogeneous catalysis and the differences between them. The similarities and differences between heterogeneous catalysis and electrocatalysis are presented both from the point of view of the materials used and the processes that take place, highlighting their applications of interest. Thus, materials with electrocatalytic activity are presented both from the point of view of the catalytic component and that with conductive properties (electrolyte). Their association in electrocatalytic processes is exemplified for various applications of interest. Among these are fuel cells and for the comparison with heterogeneous catalysis, fuel cells with solid electrolyte were presented as the most promising solid-state electrochemical systems. The role of catalysis in the study and development of electrocatalysis and the role of electrochemistry in catalysis have led to the synergistic development of both fields as well as their applications of interest for the future.

2. Heterogeneous catalysts as materials for electrocatalysis

A large number of materials employed in electrocatalysis are comparable, at the macroscopic level, to those used in catalysis, and the metals exhibiting the highest activity are frequently the same as those utilized in heterogeneous catalysis. Electrocatalysts thus contain supported metal particles, and the metal-support interaction influences their performance and stability. The resulting active sites at the metal-support or metal oxide-support interface may have specific characteristics that promote proton and electron transfer. In electrocatalytic processes, such materials serve to reduce the overpotential typically required in conventional electrochemical processes and to improve reaction kinetics. This is the key advantage of electrocatalytic processes over standard electrochemical ones [18]. The selection of electrocatalyst materials depends on the specific application and requires careful analysis of their composition, structure, and physicochemical properties to optimize electrocatalytic performance. Although a wide variety of materials have been explored historically, recent research has focused on nanomaterials and nanocomposites with high electrocatalytic activity and stability. These materials mostly contain oxide components, metals or metal alloys. Among the materials used as electrocatalysts, we can list: mixture of TiO_2-CuO mixture [19], CuO-modified TiO_2 nanotubes [20], CeO_2/ZnO [21], Cu/γ-Al_2O_3 [22], graphene-based materials (Fe, Cr-ZnO/g-C_3N_4) [23], AgNiCu nanoparticles supported on g-C_3N_4 [24], g-C_3N_4, Fe_2O_3/g-C_3N_4, Fe_2O_3/g-C_3N_4@Ni foam [25], carbon-based materials (MnY/carbon nanosheets) [26], Cu-MOF/carbon nanotubes [27], VN/Fe_3C@NCNT [28], Ni-Mo_2C/MWCNT and Co-Mo2C/MWCNT [29], NiS@MnS/multi-walled carbon nanotubes [30], layered double hydroxides (LDH) and derived structures (Ag-CoMo LDH/NF [31, 32], Fe-$NiCo_2O_4$ [33]), alloys and metallic materials (Au thin films onto nitinol substrates-NiTi [34], Al and Zn - Nb_2O_5 [35]), heterojunctions and advanced composite materials (CoNiSe and CoSe/NiSe, CoO/$NiSe_2$/NF [14], 3DOMM-TiO_{2-x} heterojunction [36]), or organometallic complexes and macrocycles (Sn(IV) 5,15-(bis-pentafluorophenyl)-10-(nitrophenyl) corroles) [37] and cucurbit[6]uril modified gold nanoparticles [38]. These materials have proven to possess superior electrocatalytic activity in reactions including electrocatalytic hydrogen evolution reactions (HER), oxygen evolution reaction (OER), oxygen reduction reaction (ORR), electrocatalytic methanol oxidation reaction (MOR), electrocatalytic CO_2 reduction reaction (CO_2RR), electrocatalytic nitrogen reduction reaction (ENRR), electrocatalytic reduction of nitric oxide, and electrooxidation of biomass-derived compounds. A detailed classification of each type of electrocatalytic material along with its specific applications is presented in **Table 1**.

Type of materials		**Application**	**Ref.**
Catalyst	**Electrolyte**		
Sn(IV) 5,15-(bis-pentafluorophenyl)-10-(nitrophenyl) corroles	0.10 M TBAP TBAP = tetrabutylammonium perchlorate	Electrocatalytic hydrogen evolution reactions	[37]
TiO_2-CuO mixture	1 M $LiPF_6$	Anode materials of lithium ion batteries (LIBs)	[19]
Ag-CoMo LDH/NF nanostructure	1 M KOH	Oxygen evolution reaction	[31]
Co_3O_4@NC	1 M KOH	Electrooxidation of biomass-based 5-hydroxymethylfurfural	[39]
AgNiCu alloy/g-C_3N_4	1.0 M KOH	Oxygen evolution reaction	[24]
Fe-$NiCo_2O_4$	1.0 M KOH	Oxygen evolution reaction	[33]
$LnFeO_3$perovskites (Ln = La, Sm, Gd) catalysts	0.1 M KOH	Oxygen reduction reaction	[40]
CoNiSe CoSe/NiSe	0.5 M H_2SO_4 1 M NaOH	Electrocatalytic hydrogen evolution reaction Photoelectrochemical (PEC) water splitting reaction	[14]

Type of materials		Application	Ref.
Catalyst	**Electrolyte**		
MnY/g-C_3N_4-derived carbon nanosheets	0.1 M KOH	Oxygen evolution reaction	[26]
Au thin films onto NiTi alloy	0.5 M H_2SO_4	Hydrogen evolution reaction	[34]
Au-CB [6] NPs CB [6]=cucurbit[6]uril	0.5 M $KHCO_3$	Electrcatalytic CO_2 reduction reaction	[38]
Cu- MOF@CNT	0.1 M KOH	Electrochemical conversion of 4-nitrophenol pollutant	[27]
CeO_2/ZnO heterojunction	0.1 mol/L Li_2SO_4	Electrocatalytic nitrogen reduction reaction	[21]
CuO-modified TiO_2 nanotubes	H_20 + methyl violet	Photodegradation and photoelectrochemical degradation of methyl violet	[20]
Cu/γ-Al_2O_3	Alkaline electrolyte	Electrocatalytic methanol oxidation reaction	[22]
VN/Fe_3C@NCNT VN = vanadium nitride	1.0 M KOH 0.1 M KOH	Hydrogen evolution reaction Oxygen evolution reaction And oxygen reduction reaction	[28]
CoO/$NiSe_2$/NF NF = nickel foam	1 M KOH	Electrocatalytic water splitting (HER and OER)	[41]
Al and Zn—Nb_2O_5	0.5 M H_2SO_4	Electrocatalytic hydrogen evolution reaction	[35]
(Fe, Cr)-ZnO/g-C_3N_4 nanocomposite	1 M KOH	Electrocatalytic water splitting (OER and HER.)	[23]
g-C_3N_4, Fe_2O_3, Fe_2O_3/g-C_3N_4, Fe_2O_3/g-C_3N_4@NF NF = nickel foam	1 M KOH	Oxygen evolution reaction	[25]
Ni-Mo_2C/MWCNT Co-Mo_2C/MWCNT	Mixture of 1 M KOH and 1 M CH_3OH	Electrocatalytic methanol oxidation reaction	[29]
NiS@MnS/MWCNT MWCNTs = multi-walled carbon nanotubes	1 M NaOH	Electrocatalytic methanol oxidation reaction (MOR)	[30]
0D/3D g-C_3N_4 quantum dots/3DOMM-TiO_{2-x} heterojunction	0.1 M K_2SO_4	Electrocatalytic reduction of nitric oxide	[36]

Table 1.
Overview of electrocatalytic materials and their corresponding applications.

Most electrocatalytic materials are heterostructures, whose combined properties produce a synergistic effect on electrocatalytic performance. The type of each component, their mutual interactions, as well as the type of the support and its interaction with the components, is all important factors that govern both the electrochemical and the catalytic processes. Therefore, the construction of heterojunctions represents an effective strategy for developing highly active electrocatalytic materials. For example, Wu et al. [31] proposed the synthesis of a hierarchical nanostructure composed of silver-modified cobalt-molybdenum layered double hydroxides (Ag-CoMo

LDH/NF) for oxygen evolution reaction. The incorporation of silver facilitated the formation of well-defined heterostructures with the Co/Mo LDH matrix and affects the local electron density [31]. Moreover, heterostructured architecture can enhance the concentration of available active sites and improved charge transfer efficiency, ultimately boosting the oxygen evolution reaction [32]. In this case, the improved properties of the electrocatalyst also come from the excellent corrosion resistance conferred by the presence of Mo. Molybdenum also contributes to the generation of porosity, thus expanding the interface area with the electrolyte, and facilitating the generation of active catalytic sites.

Li et al. [21] reported the synthesis of a CeO_2/ZnO heterojunction exhibiting outstanding performance in the ENRR reaction, achieving a high ammonia yield of 60.21 μg h^{-1} mg cat^{-1} and a Faradaic efficiency of 11.48% at an applied potential of −0.2 V versus RHE, using 0.1 mol L^{-1} Li_2SO_4 as the electrolyte. Although the CeO_2/ZnO heterostructure exhibits a lower electrochemically active surface area (ECSA) than its individual components (CeO_2 and ZnO), it demonstrates superior ENRR performance. This behavior is primarily attributed to the catalytic effect induced by the heterostructure, particularly the "electron pump" mechanism that enhances charge transfer and facilitates nitrogen activation. These findings indicate that, in this case, ECSA is not the decisive factor governing catalytic activity in the ENRR. Another type of heterostructure was developed by Alhalili's group [41], consisting of CoO/$NiSe_2$ supported on nickel foam, which exhibited remarkable performance in overall water splitting (HER and OER). The improved electrocatalytic activity is ascribed to the synergistic effect of CoO and $NiSe_2$. CoO effectively modulates the electronic structure of $NiSe_2$/NF, increases the electrochemically active surface area (ECSA), and improves reaction kinetics. CoO/$NiSe_2$ heterojunction generates a higher density of active sites and facilitates charge transfer, thereby significantly boosting electrocatalytic efficiency.

Iron oxide (Fe_2O_3) combined with g-C_3N_4 and nickel foam (FGCN@NF) was reported by Kumaravel et al. [25] to form a highly efficient electrocatalyst for photocatalysis and the oxygen evolution reaction (OER). The enhanced electrocatalytic performance of the heterostructure is attributed to the strong interaction between Fe_2O_3 and g-C_3N_4, along with an enhanced number of electrochemically active sites. Additionally, the synergistic effect between the components promotes electron transfer and lowers the energy barrier for proton adsorption and desorption, thereby improving OER activity.

In an electrocatalytic process, in addition to the electrochemical effects described above (such as electronic modulation at the interface of heterostructure components and changes in electron transfer), structural, textural, and morphological properties of composite materials are also of great significance. In the study reported by Yang et al. [14], the positive influence of cobalt vacancies, introduced using ethylenediaminetetraacetic acid (EDTA) during the synthesis of a CoSe/NiSe heterostructure, was demonstrated using photoelectrochemical (PEC) water splitting. The vacancy-containing sample exhibited the lowest overpotential compared to the vacancy-free counterpart, indicating that the presence of cobalt vacancies enhances the density of active catalytic sites and promotes charge transfer, ultimately improving the electrocatalytic performance. The effect of vacancies within the 0D/3D g-C_3N_4 quantum dots/3DOMM-TiO_{2-x} heterojunction structure on the electrocatalytic reduction of nitric oxide was also investigated by Chen et al. [36]. The presence of oxygen vacancies and Ti^{3+} species was found to be critical, as these defects enhance gas adsorption and activation on the catalyst surface and increase the density of active sites on TiO_2.

The heterojunction between g-C_3N_4 quantum dots and 3DOMM-TiO_{2-x} improves electrical conductivity, facilitating photogenerated electron transfer and reducing interfacial charge-transfer resistance at the electrode-electrolyte interface. Additionally, the high specific surface area of the composite promotes stronger interaction with NO molecules and exposes a greater number of active catalytic sites. Density functional theory (DFT) calculations confirmed that NO adsorption on the surface of g-C_3N_4 quantum dots is thermodynamically favorable, as evidenced by negative Gibbs-free energy values.

The support material employed for the immobilization of catalytically active species in electrocatalytic applications plays a critical role and strongly affects the overall efficiency of the process. Recently, Ebrahimi et al. [30] reported the synthesis of a new material based on NiS@MnS supported on multi-walled carbon nanotubes (MWCNTs) for the electrocatalytic methanol oxidation reaction (MOR). In their study, MWCNTs were not found to participate directly in the oxidation of methanol; instead, their role was attributed to support-related effects, facilitating methanol adsorption by enhancing catalyst dispersion, increasing the specific surface area and electrical conductivity, and providing more accessible active sites. The nanorod morphology of the support plays a critical role in the electrocatalytic process. It ensures the exposure of a greater number of catalytically active centers and facilitates electron and ion transfer. This behavior ultimately leads to enhanced catalytic activity for methanol oxidation. In addition to morphology, the specific surface area of the material used in electrocatalysis is a key property. Through specific methods, materials can be engineered to possess increased surface areas, which are beneficial in electrocatalytic processes. For instance, Heydari-Bafrooei et al. [35] demonstrated that doping Nb_2O_5 with Zn and Al not only enhances kinetics of electron transfer by altering its electronic structure but also improves the surface area and morphology. This results in more accessible sites for the hydrogen evolution reaction (HER) and consequently boosts the electrocatalytic activity.

Superior electrocatalytic activity in the oxygen evolution reaction (OER) was demonstrated by Wei et al. [33] for Fe-$NiCo_2O_4$ calcined at 350°C and was attributed to the improved textural properties (significant surface area, highly porous structure), as well as the structural features obtained (unique amorphous/crystalline structure). These characteristics allow for the exposure of more accessible active sites to the reaction, thereby favoring the electrocatalytic process. This highlights the key role of the synthesized material and the applied treatments (such as calcination at various temperatures) on the electrocatalytic reaction.

3. Materials for solid oxide fuel cell electrodes

Solid oxide fuel cells (SOFCs) are important as electrocatalytic devices because they directly convert chemical energy into electricity with high efficiency and low emissions, using electrocatalysis to drive the reactions. The performance of an SOFC depends largely on the electrocatalytic activity of its electrodes. The main property of an SOFC electrode is the ability to catalyze the corresponding electrochemical reaction while allowing the released electrons to reach the current collectors. Therefore, the material used in this application must possess certain properties, of which mixed conductivity (electronic and ionic) and stability under SOFC operating conditions (high temperatures, reducing/oxidizing atmosphere) play an important role [42]. The catalytic processes in SOFCs involve the anode, which facilitate fuel conversion, and the cathode, which accelerates oxygen reduction. The development of new materials and architectures to enhance the oxygen reduction reaction at the cathode and the fuel oxidation reaction at the anode will continue to have a major impact on the development of fuel cells.

Nanostructured materials created by synthetic design have found great utility in all aspects of human activity, one of them is on the energy and fuels. Yttria-stabilized zirconia (YSZ) is among the most effective materials used as an electrolyte or anode component in SOFCs due to its excellent ionic conductivity, chemical stability, and superior mechanical strength. New properties were evidenced for nanometer-sized YSZ particles. Lattice defects introduced by the ionic substitution of Y^{3+} for Zr^{4+}can improve the redox capability as well as the activity and mobility of lattice oxygen. Mesoporous zirconia-yttria and ceria-zirconia-yttria with large surface area have been synthesized by the hydrothermal method from simple inorganic salts, instead of expensive alkoxides, in aqueous medium with or without surfactants [43–47]. The cermet structure of a solid oxide fuel cell, most commonly Ni-YSZ (yttrium-stabilized zirconium), provides a number of important functions, such as adhesion of electro-catalytically active species (Ni), ion conductivity, porosity for gas transport, and high thermal conductivity. Many material properties observed in nanotechnology arise from solids having large surface areas and uniform, high porosity at the nanometer scale. In nanoparticles, the ratio of surface atoms to bulk atoms is significantly higher than in micrometer-sized materials, which can result in unique properties. TiO_2-Y_2O_3-ZrO_2 (TYZ) mixed oxides were used in preparation of the composite anodes for low-temperature solid oxide fuel cell applications. The influence of various synthesis parameters deemed important for mesostructure formation, such as method, surfactant, molar composition, mixing temperature, and aging temperature on the structure, morphology, electrochemical, and catalytic properties, has been reported. One of the main challenges in the synthesis of materials for SOFCs is the synthesis of intermediate temperature active materials (IT-SOFC). For this purpose, the effect of both the composition and the synthesis method on the catalytic and electrochemical activity of the obtained materials was studied.

The first studies were based on the CeO_2–YSZ system, described by Gorte et al. [48], which was a real success. Thus, the control of the porosity of the obtained oxide nanomaterials as well as their electrochemical and catalytic properties was sought. In the case of electrodes, where the electrocatalytic processes take place, porosity plays an important role in allowing the diffusion in the gas phase of the fuel at the three-phase boundary (TPB) in SOFC and the high surface area of the nanoparticles to ensure the optimal conditions for the heterogeneous catalytic processes that take place. Many studies have focused on the electrochemical and catalytic behaviors of anode-based ceramics, for example, Y-doped La0.7Sr0.3CrO3, yttrium-doped $SrTiO_3$, Sr_2MgMoO_6, $Sr_{0.88}Y_{0.08}TiO_3$–CeO_2, $Ce_{0.8}Ca_{0.2}O_{1.8}$, Ce-doped LaSr chromo-manganite, La-doped $SrTiO_3$, CeO_2–Y_2O_3–ZrO_2 [49–52]. Practically, the synthesis methods require techniques such as solid-state reactions, coprecipitation, or impregnation. CeO_2 is well known as a true "oxygen reservoir," which is why multiple studies have been carried out on (CeO_2)-YSZ ceramic materials with mesoporous structure that have been synthesized by direct synthesis using surfactant self-assembly [51]. In these reactions, the ordered surfactant micelles serve as a template for the condensation of inorganic agents. It is well established that the Ce^{3+} dispersion, structure, and surface characteristics of metal oxides significantly influence catalytic activity. The catalytic performance of the anode is enhanced by doping the mixed oxides with transition metals such as Ni or Cu [45, 50, 53] and noble metals such as Pd, Pt, and Rh [54], which significantly increases the performance of the cells, especially when using CH_4 as fuel. Copper-ceria-yttria-stabilized zirconia material used as anode combines the high electronic conductivity of copper and ceria with the structural stability of YSZ, making it a promising alternative to the more common Ni-YSZ anodes. The use of Cu-ceria-YSZ anodes has been shown to achieve reasonable performance over a wide range of fuels, even using synthetic diesel fuel. These materials used as electrocatalysts in fuel cells have both catalytic and conductive properties. In **Figure 1**, the ionic conductivity and catalytic activity of some oxide nanomaterials ($CeO_{2-x}PrO_2$, where x = 10%-CP10, 50%- CP50, 90%-CP90) with variable content of Pr are presented comparatively. Thus, an effect of the composition is observed both on the propane conversion and on the ionic conductivity. For example, if in the case of conversion a significant increase is observed between 5 and 10% PrO_2 concentration, in the case of conversion a more significant increase is observed between 10 and 50% PrO_2. These variations are due both to the concentration of Ce^{3+} ions, with a significant effect on the catalytic activity, and to the reduction in the size of the mixed oxide particles with increasing Pr content. XPS results showed that for a percentage greater than 5% PrO_2 a considerable increase in the Ce^{3+} content is observed, while above 10% PrO_2 the share of Ce^{3+} oxidation states remains approximately constant. These mixed oxides were obtained by the surfactant (cetyltrimetylammonium bromide-CTAB)-assisted sol-gel method and hydrothermal treatment [55]. One of the methods used for the synthesis of the materials for SOFC applications in which sol transforms into gel is the so-called Pechini process [56]. Using this method, oxide mixtures $(Ce_{0.9}(Eu_{1-x}Sr_x)_{0.1}O_{2-\delta}$ (x = 0, 0.1, 0.3, 0.5, and 0.7) were prepared [57] as electrolyte material. Impedance spectra reveal a notable reduction in grain boundary resistance following the partial substitution of europium with strontium in europia-doped ceria. Consequently, the incorporation of strontium into the europium-doped ceria matrix led to an increase in total conductivity.

Zirconium-based materials such as yttria-stabilized zirconia (YSZ), gadolinium oxide-doped ceria (GDC) materials, and lanthanum gallate-based materials such as $La_{0.8}Sr_{0.2}Ga_{0.8}Mg_{0.2}O_{3-\delta}$ (LSGM) are the most common choices [58] for oxygen-ion conducting solid oxide fuel cells (O–SOFC were the electrolyte), which offer pathways for oxygen ions to migrate through oxygen vacancies. In proton-conducting solid oxide fuel cells (H–SOFCs) [59] $BaCe_{0.7}Zr_{0.1}Y_{0.2}O_{3-\delta}$ (BCZY) is the most

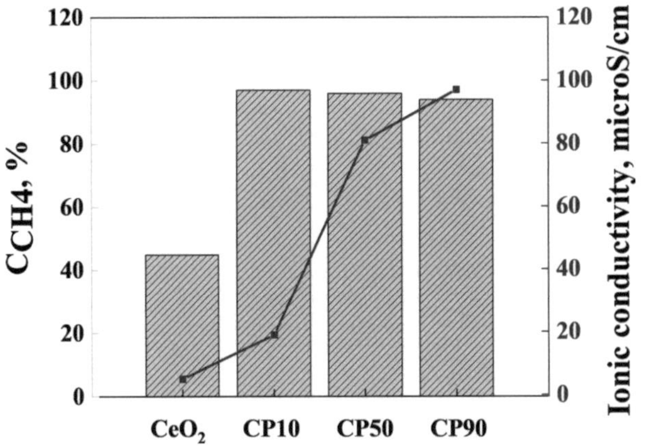

Figure 1.
Variation of catalytic activity and ionic conductivity with praseodymium content (Unpublished).

commonly used electrolyte. Compared with O-SOFCs, H-SOFCs are more commonly employed at intermediate and low temperatures because of their superior performance. These are due to the activation energy of proton transport, much lower than that of oxygen ion transport. The electroactivity of the anode can be modified by replacing the pure oxide-ionic conductor of the cermet with a proton-conducting ceramic, thus favoring hydrogen absorption and higher proton mobility. Cermet anodes made of Ni-$CaZr_{0.95}Y_{0.05}O_{2.975}$ (Ni-CZY) and Ni-$SrZr_{0.95}Y_{0.05}O_{2.975}$ (Ni-SZY) are thus obtained [60].

Perovskites are also utilized in solid oxide fuel cells (SOFCs). Perovskite oxides have been investigated as alternatives to cerium-based systems because of their high ionic and electronic conductivity and stability under the reducing conditions [61]. The most studied perovskites are titanates and chromites. Also, the double perovskite Sr_2MgMoO_6 is a promising anode material, demonstrating long-term stability, high sulfur resistance, and excellent performance. Lanthanum-cobaltite-based perovskite oxides are the representative mobile conductors that have been extensively investigated as the potential cathodes for SOFCs. $La_{1-x}Sr_xCoO_{3-d}$ (LSC) perovskites demonstrate the favorable oxygen-ionic conductivity alongside high electronic conductivity. Perovskites can also function as electrolytes. Notably, high oxygen ion conductivity has been observed in perovskite-structured oxides derived from lanthanum gallate ($LaGaO_3$), particularly within the $La_{1-x}Sr_xGa_{1-y}Mg_yO_{3-d}$ (LSGM) series. Partial substitution of strontium by lanthanum and the incorporation of divalent magnesium cations into the gallium sublattice enhance the concentration of oxygen vacancies in the material. Additionally, doping with transition metals also enhances the electrical conductivity of LSGM.

The electrochemical performance of cathodes is influenced not only by their chemical composition but also on their microstructure characteristics. The microstructure of the cathode is largely related to the deposition process applied. La_{1-x}Srx-MnO_3 perovskite (LSM) has been frequently used as a cathode material for SOFCs due to its high catalytic activity in oxygen reduction and compatibility with yttrium-stabilized zirconia (YSZ). Although LSM has demonstrated high performance for SOFCs operating at temperatures of approximately 800°C, it significantly decreases with the operating temperature. For this reason, studies have focused on cobalt-containing perovskite oxides (e.g., $Sm_{0.5}Sr_{0.5}CoO_3$-SSC), which exhibit mixed conduction and comparatively greater ionic conductivities than LSM owing to an increased concentration of oxygen vacancies [62, 63].

$LiNi_{0.8}Co_{0.15}Al_{0.05}O_{2-\delta}$ (NCAL) has been extensively applied in low-temperature hydrogen-based SOFCs [64]. This material has a double catalytic activity for both the oxygen reduction reaction and the hydrogen oxidation reaction. At the same time, the $H^+/O^{2-}/e^-$ conductance is triple. The application of NCAL as an electrode for direct ethanol SOFCs has improved the performance of these fuel cells.

Among the anode materials developed in the recent years are supports based on nanostructured conductive polymers, which have offered the possibility of large-scale commercialization of direct alcohol fuel cells [65]. In many situation, the polymer conductor is combined with carbon materials used as substrate as nanotube, nanofilm, or three-dimensional structure.

4. Heterogeneous catalysis and electrocatalytic processes in fuel cells

The exhaustion of fossil fuels and the environmental contamination resulting from rapid population growth and industrial development have stimulated research into new technologies for clean and sustainable energy generation and storage, such as fuel cells and solar cells. Advances in the field of renewable energy have demonstrated that fuel cells—because of their high efficiency and versatility in using different fuels—represent a promising energy source with a wide range of potential applications. The global interest in accessing cleaner electricity has encouraged the adoption of renewable alternatives and the implementation of efficient energy production methods with the minimal environmental impact. As a result, fuel cells are regarded as a powerful and attractive means of generating energy through electrochemical and catalytic processes.

The main redox reactions from electrodes of these renewable energy technologies are the oxygen reduction reaction (ORR), the oxygen evolution reaction (OER), the hydrogen evolution reaction (HER) and the CO_2 reduction reaction (CO_2RR) [66–71]. All of these reactions require materials with catalytic properties. Due to the disadvantages of catalysts based on noble metals such as iridium, palladium, platinum, with limited availability, poor stability, limited selectivity and high costs, considerable efforts have been made to develop a wide range of functional or multifunctional electrocatalysts without noble metals. Interest has been growing in the recent years in the use of hydrogen as a fuel in power generation.

Fuel cells have been categorized according to the materials employed for their electrolytes. The proton exchange membrane fuel cells (PEMFC) work at lower temperature (60–80°C), the electrolyte is based on a polymer, and the fuel is H_2 and electrical efficiency 45–55%. The alkaline fuel cells (AFC), phosphoric acid fuel cells (PAFC), direct methanol fuel cells (DMFC), and molten carbonate fuel cells (MCFC) operate at temperatures between 140 and 200°C, using hydrogen, methanol and even hydrocarbons (MCFC) as fuel and efficiency 40–55%. The work temperature of solid oxide fuel cells (SOFC) can be between 500 and 1000°C, and their electrical efficiency is between 50 and 60%. The first have applications in transportation, military, and public facilities and the latter for power generation and hybrid applications [72].

The development of bio- and photo/electrocatalysts was conceived to simultaneously improve power generation and pollutant removal in microbial fuel cells (MFC) and photo electrochemical cells (PEC). Microbial fuel cells have been regarded as the most effective approach for wastewater treatment, as they can employ bacteria as biocatalysts to generate electricity and oxidize organic matter [73–75]. The highest reported power density has reached 1000 W/m^3. MFCs have also disadvantages, such as complex operation, bacterial cultivation, long start-up time, and strict working conditions. As an example, to enhance the efficiency of organic pollutant degradation in water using a photo-assisted microbial cell, the Ag-mIm/0.01 g-C_3N_4 composite—characterized by strong visible light absorption—was employed as the cathode material [75]. It was considered that the cell performance is the result of the synergistic effects of the photocathode and bioanode, which simultaneously generates electrons as reducing species and OH as oxidizing agent. The Ag-mIm/0.01 g-C_3N_4 composite demonstrated a strong absorption of visible light.

Photoelectric cells have several advantages compared to conventional fuel cells. Their most important advantage is that they can operate with any organic material under ambient conditions, without any other energy input than light [76, 77]. Their disadvantage is the need to absorb light at a specific wavelength corresponding to the photocatalyst used. Many biomass and biomaterials, such as polysaccharides, proteins, cellulose, lignin, saccharides, amino acids, alcohols, organic acids, urea, from water could be photodecomposed by photoanode coupled with an O_2-reducing cathode. In the presence of light, a photocatalytic process occurs in which electrons are promoted from the valence band to the conduction band of the photosensitive material employed as an anode, thus generating electron/hole pairs. Positive holes are the strong oxidants for the degradation of organic compounds into CO_2 and H_2O, and negative electrons are the strong reducers. The critical factor determining the efficiency of photocatalytic degradation comprises in obtaining and using photocatalytic materials active in visible light, which determines the separation of photogenerated charge carriers, preventing their recombination. At the same time, an effective method to separate photogenerated pairs (e^-, h^+) is the electrochemical pathway. This can be accomplished in a PEC cell comprising two electrodes: a photoanode consisting of the photocatalyst and a cathode (Pt). The electrolyte consists of water, the photodegradable water-soluble compound, and a salt, which is required to ensure ionic conductivity. A wide variety of materials have been synthesized and used in PEC applications. The materials used to make the anode, where the photocatalytic process takes place, are semiconducting oxides or composite materials that have both photocatalytic activity and electrical conductivity. The vast majority of them contain TiO_2, which is the most widely used photocatalyst [78, 79].

PEC can be associated with a photoelectrochemical cell in a composite system. This composite system contains a photoelectrocatalytic reactor and a photocatalytic combustion cell, thus increasing the degradation efficiency of organic compounds [79, 80].

In the presence of oxygen, no external electrical polarization between the electrodes is required, since in this case a potential difference of a few hundred mV between the two electrodes could be obtained without much effort. Thus, photocells, which operate in the presence of oxygen, are spontaneously operating devices. The overall effect of the system is the degradation of organic compounds through the photocatalytic process, simultaneously generating electrical energy. The efficiency of a photovoltaic device is simply the ratio of the electrical energy produced by the device to the light power incident on the device.

The photoelectrochemical properties of the mesoporous Ce-modified TiO_2 photocatalysts CeO_2 (TGP, TGP2.5C, TGP5C with 0, 2.5, and 5% Ce) were evidenced [79] by comparing their photovoltaic (**Figure 2a**) and photocatalytic (**Figure 2b**) properties.

The photovoltaic activity was evaluated by measuring the current density-voltage curves under the simulated solar light irradiation. For all the analyzed photoanodes, the

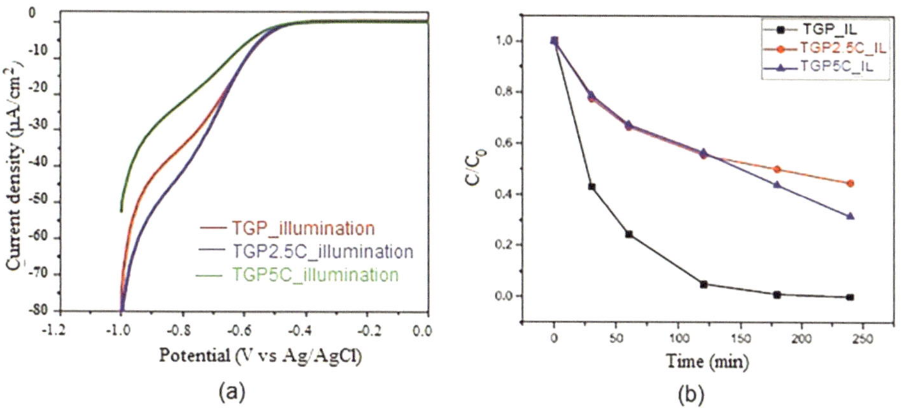

Figure 2.
The variation of current density as a function of electrode potential measured vs. Ag/AgCl for Ce-modified TiO_2 samples (a) and photocatalytic degradation of phenol (b) under simulated solar light irradiation. Reprinted from Ref. [79] with permission from Elsevier.

photocurrent density increased under illumination, confirming their semiconductor behavior. The photocatalytic degradation of phenol in aqueous medium was evaluated also in visible light. Electrochemical evaluation of photoactivity that the Ce-modified mesoporous TiO_2 photocatalysts could be considered for degradation of organic compounds from wastewaters by advanced oxidation processes (AOP) or for photoanodes construction as the efficient photoelectrochemical fuel cell (PFC) systems.

Solid oxide fuel cells (SOFCs) have attracted attention due to their ability to convert chemical energy into electricity without intermediates [67–71]. However, they have the disadvantage of high operating temperatures that can affect the properties of the materials used. The lowering their operating temperatures can result in reduced catalytic activity at both the anode and cathode. Therefore, research has focused on developing new anode and cathode materials that are more active at moderate temperatures. The generation of various interfaces can improve the triple phase boundary (TPB) of the material, increase the number of operational sites, thereby optimizing the efficiency of the electrode, and strengthen the durability of the cell.

A solid oxide fuel cell (SOFC) comprises an anode, an electrolyte, and a cathode [69]. Fuel is delivered to the anode side, while oxygen (from air) enters *via* the cathode. The operation of the SOFC relies on the fuel oxidation reaction at the anode and the oxygen reduction reaction at the cathode. The fuel most commonly used is hydrogen, although other fuels such as glycerol, propane, methanol, ethanol, and ammonia can also be utilized. In oxygen-conducting SOFCs, oxygen from the air is reduced at the cathode to form O^{2-} ions, which migrates through the electrolyte, a conductive oxide material. At the anode, the O^{2-} ions react with the hydrogen in the fuel to form water, releasing two electrons into an external circuit [80–82]. The flow of electrons from the anode to the cathode generates an electric current. In this way, the chemical energy from hydrogen combustion is efficiently converted into electrical energy through a clean electrochemical process.

Among fuel cells, solid oxide fuel cells (SOFCs) are by far the most efficient (60%) for converting hydrogen, or another fuel, directly into electricity and one of the cleanest routes for electricity generation, with ammonia emerging as a promising zero-carbon alternative fuel.

In recent years have been intensively studied direct ammonia solid oxide fuel cells (DA-SOFCs) as an efficient ammonia-fueled power system. DA-SOFCs hold significant potential for application in the transportation sector due to their low greenhouse gas emissions. Unfortunately, incomplete decomposition of ammonia can lead to emissions of NO_x (nitrogen oxides) that are air pollutants. Research efforts are focused on minimizing NO_x formation and developing effective emission control strategies. Currently, ammonia-fueled solid oxide fuel cells utilizing Ni-based catalysts have achieved peak power densities of over 1000 mW cm^{-2} [83]. In DA-SOFCs, the ammonia decomposition reaction (ADR) and hydrogen oxidation reaction (HOR) are crucial for efficient power generation. ADR breaks down ammonia into hydrogen and nitrogen, while HOR oxidizes the generated hydrogen, producing electricity. A key challenge in DA-SOFCs is optimizing these reactions to maximize performance, especially at lower temperatures. The rate of ADR is strongly influenced by temperature, with elevated temperatures typically resulting in faster decomposition. HOR is an exothermic reaction that occurs at the anode, where hydrogen reacts with oxide ions (O^{2-}) to produce water and electrons. The electric current is thus generated in the fuel cell. This reaction from anode can be used to drive the endothermic ADR, creating an efficient heat integration process. HOR is usually a faster reaction than ADR, and its kinetics is well understood in SOFCs [84–86].

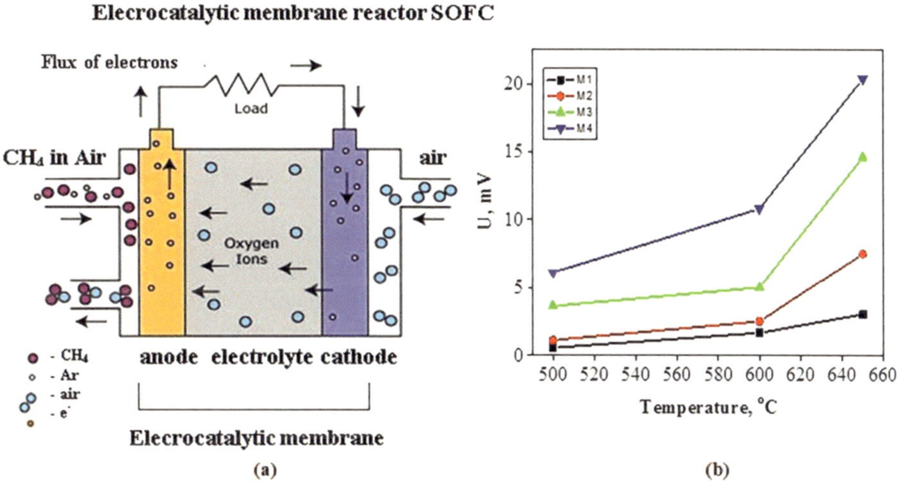

Figure 3.
Scheme of the electrocatalytic membrane reactor (a) and the variation of its performances with membrane composition and reaction temperature (b) (Unpublished).

Membrane catalytic reactors have developed as an interdisciplinary field at the border between catalysis, materials science, and membrane processes. The membrane does play not only the role of a separator, but also a catalyst being part of the reactor. Thus, membrane reactors with the role of a fuel cell were realized [87]. The electrochemical membrane reactor was used for direct oxidation of hydrocarbons [88, 89] and was considered an alternative, as catalytic tests of electrodes, for solid-oxide fuel cell (SOFC) that uses hydrocarbons directly without reforming. The development of direct-oxidation fuel cells may have significant implications for the commercialization of fuel cells because direct oxidation of hydrocarbon fuels is not limited to CH_4 [90]. SOFC fuel cells and the need to improve their performance have created conditions for associating membrane catalytic processes with electrochemical processes. Membrane for electrocatalytic reactor can be a sandwich of anode-electrolyte-cathode (**Figure 3a**).

For the membrane reactor shown in **Figure 3**, membranes with the following composition were used: M1-Anode-Ni5YSZ/Electrolyte-YSZ/Cathode-$Ce_{0.9}Pr_{0.1}O_2$; M2-anode-Ni10YSZ/electrolyte-YSZ/cathode-$Ce_{0.9}Pr_{0.1}O_2$; M3-anode Ni20YSZ/electrolyte-YSZcathode-$Ce_{0.9}Pr_{0.1}O_2$; M4-anode-Ni30YSZ/electrolyte-YSZ/cathode-$Ce_{0.9}Pr_{0.1}O_2$. In the composition of the anodic material, the percent of nickel was 5, 10, 20, and 30%. Effects of the membrane-anode composition and working temperature on the performance of membrane electrocatalytic reactor are presented in **Figure 3b**.

These results confirm that the membrane electrocatalytic reactor is a good solution for testing, through direct oxidation of hydrocarbons of the anode materials for low-temperature fuel cells.

5. Synergy and differences between heterogeneous catalysis and electrocatalysis, their influence on future technologies

Catalytic and electrocatalytic processes both accelerate chemical reactions using catalysts, but they differ in how they provide the energy for the reaction. Catalysis uses chemical energy to lower activation energy, while electrocatalysis utilizes electrical energy to achieve the same. Electrocatalysis and heterogeneous catalysis involve elementary bond-breaking and bond-forming processes in chemical reactions, sharing common mechanistic principles. Although there are many elements of synergy and similarities in materials used and reaction mechanisms, there are also some differences between these processes [5, 91, 92]. A summary of these is presented in **Table 2**.

Heterogeneous catalysis enables a more detailed investigation of reaction mechanisms, particularly for gas-phase reactions. This advantage arises from the relatively straight forward characterization of the catalyst surface and the gas/solid interface, in contrast to the more complex aqueous/solid interface encountered in electrocatalysis. As a result, heterogeneous catalysis often provides deeper mechanistic insights, supported by spectroscopic and theoretical analyses of the gas/solid interface, whereas the inherent complexity of the aqueous/solid interface poses greater challenges for obtaining comparable information in electrocatalytic systems.

Category	**Differences**		**Similarities**
	Heterogeneous catalysis	**Electrocatalysis**	
Energy source	Driven by the chemical potential of reactants	Driven by electrical energy applied to the electrode	Both accelerate reactions by lowering activation energy
Activation energy reduction	Achieved by providing an alternative reaction pathway with a lower energy barrier	Achieved by facilitating electron-transfer reactions at the electrode surface	Both modify the reaction pathway to make it energetically favorable
Reaction environment	Reactions occur mainly at the gas/solid or liquid/solid interface	Reactions occur at the solid/liquid interface (electrode-electrolyte)	Both rely on surface reactions at the catalyst interface
Interface complexity	The gas/solid interface is relatively simple and allows detailed mechanistic study	The aqueous/solid interface is more complex due to ions, electric fields, and solvents	Insights from heterogeneous catalysis help understand electrocatalysis
Charge and mass transfer	Involves neutral species; no direct electron flow through an external circuit	Involves ion, electron, and proton transfer through the electrolyte and electrode	Both involve adsorption, reaction, and desorption at active sites
Phases involved	Two phases: solid catalyst and reactant (gas or liquid)	Two or three phases: electrode (solid), electrolyte (solid or liquid) and reactant (gas or liquid)	Three-phase boundary (TPB) represents the point of synergy between catalytic and electrochemical processes [93–95]
Applications	Hydrocarbon oxidation, industrial catalytic processes	Fuel cells (SOFC, PEMFC), electrolysis, redox reactions	Both contribute to energy conversion and environmental technologies
Key performance factors	Surface composition, morphology, and active site availability	TPB density, electrode microstructure, ionic/electronic conductivity	Optimizing surface structure and contact enhances performance in both

Table 2.
Similarities and differences between heterogeneous catalysis and electrocatalysis.

Electrocatalysis has frequently relied on insights provided by heterogeneous catalysis in the gas phase. The growing interest in novel applications of heterogeneous catalysis for energy conversion, particularly involving catalytic transformations at the fluid/solid interface has driven the development of electrocatalytic approaches.

Comparing catalytic and electrocatalytic processes, it can be highlighted that both catalysts and electrocatalysts lower the activation energy of a reaction. The reduction of activation energy is achieved, in the case of catalysis, by providing an alternative reaction pathway with a lower energy barrier, and in the case of electrocatalysis by facilitating electron transfer reactions at the surface of an electrode. In the case of catalysis, the energy for the reaction is derived from the chemical potential of the reactants themselves, while in the case of electrocatalysis, the energy is obtained from the electrical energy that is applied to the electrode, influencing the kinetics of the reaction and allowing reactions that otherwise could not occur. The important differences between catalysis and electrocatalysis result from the presence of solution, counterions, and electric fields. It is believed that understanding the similarities and differences between molecular transformations occurring under ultra-high vacuum and electrochemical conditions will drive progress in the development of catalytic and electrocatalytic materials and processes.

Particularly in the electrocatalytic processes that take place in solid oxide fuel cells (SOFC) and polymer electrolyte membrane fuel cells (PEMFCs), three-phase boundary (TPB) is studied. TPB is a critical location where distinct phases meet: the electrolyte, electrode (electron conducting phase), and the fuel/oxidant (gas or liquid). In this space, the synergistic action of the catalytic and electrochemical processes is evident. In SOFC, TPBs are the active sites where fuel oxidation and oxygen reduction reactions take place. The length and quality of the TPB significantly influence the fuel cell's overall performance. Increasing the TPB length can improve the reaction rate and thus the power output. In PEMFCs, the TPB is also crucial for water management. Proper hydration of the electrolyte at the TPB is essential for proton conductivity and preventing performance degradation. Designing electrode structures with high surface area and optimized pore structures can enhance TPB density and performance [96]. Using appropriate catalysts and optimizing their dispersion within the electrode can improve the TPB area and reaction kinetics. **Figure 4** shows the microstructure of an anodic material and the processes that occur at three-phase boundary in a SOFC.

As can be seen in **Figure 4**, the catalytically active sites are on the surface of the Ni-YSZ anode. The dissociative adsorption of hydrocarbon (CH_4) is favored by the metallic phase (Ni) and the oxidation by the ionic oxygen species in the mixed oxide network (in this case yttrium-stabilized zirconium dioxide modified with TiO_2) with ionic conductivity. By oxidizing the carbon species formed by methane dissociation with oxygen from the air, CO_2 is formed and the oxygen vacancies in the mixed oxide network are filled. In the membrane electrocatalytic reactor (**Figure 3**), the electron flow formed following fuel oxidation moves from the anode to the cathode. The ionic oxygen species resulting at the cathode in the presence of air migrate through the electrolyte to the anode where they perform hydrocarbon oxidation. The efficiency of the electrocatalytic process in the reactor is determined by the most intimate contact between the oxide material acting as an electrolyte and the deposited metal species, respectively the oxide material that favors the formation of ionic oxygen species. This contact can be improved by using, for example, colloidal silver.

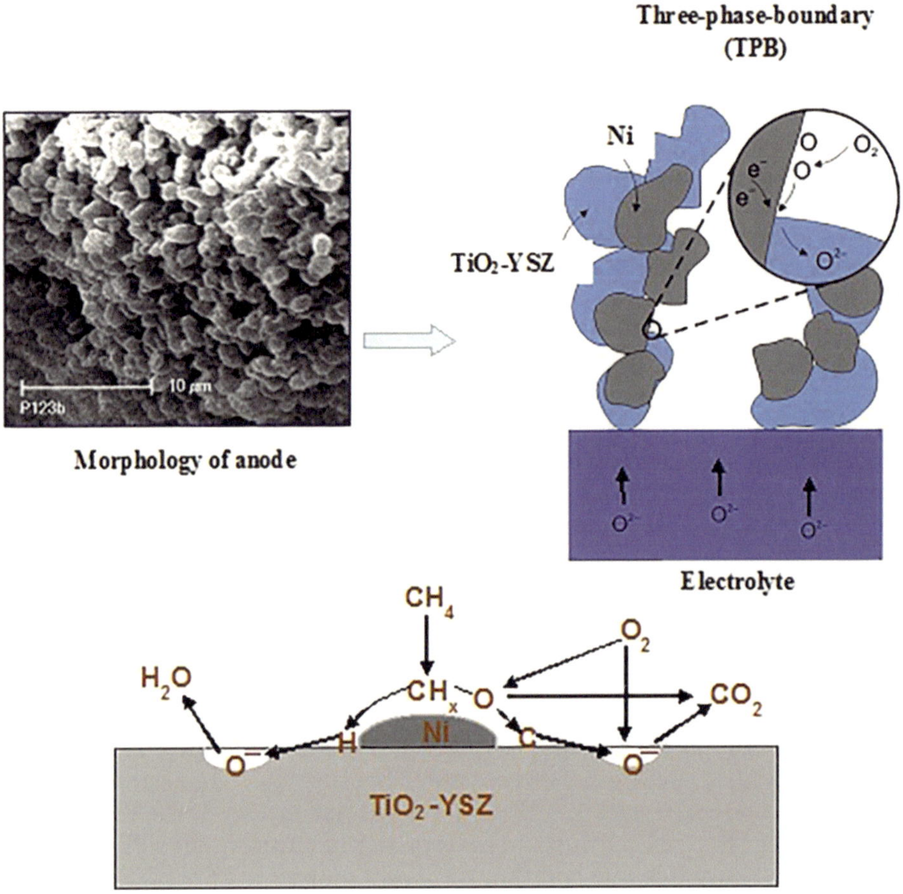

Figure 4.
Microstructure of anode material, TPB at interface with electrolyte, and the catalytic reaction at interface electrocatalyst-gas phase.

The effective TPB density in SOFCs and its relationship with a better known parameter, the thickness of the active anode, were also studied by calculation using numerical methods and the results compared with the experimental cell performances [97].

The results of this study hold substantial significance for the design and operational optimization of SOFCs. It was thus observed that both the thickness of the active anode and the effective density of the TPBs can influence the electrochemical performance of the cells, allowing the understanding of the microstructural activity in order to design more efficient and durable cells. Optimizing the TPB locations also plays a decisive role in the development of more efficient and reliable SOFCs. A theoretical model of electrospun electrodes made in comparison with fibers with surfaces covered with impregnated particles allowed to establish the key geometric parameters and their interrelation, which are necessary to obtain the key parameters of the electrode [93]. Thus, the length of the three-phase boundary (TPB) in the electrospun electrode was determined. Additionally, the influence of particle radius, fiber radius, and impregnation load was examined. The effects of particle radius, fiber radius, and impregnation charge were also studied. It was thus proven that such a mathematical model can contribute to the design of the optimization of the electrospun SOFC electrode.

To understand the effect of gas transfer on the electrochemistry at the three-phase interface, scanning electrochemical cell microscopy (SECCM) was used [95]. Thus, using ORR as an example of a reaction involving a gaseous reactant, an increase in the limiting current was obtained with the kinetics of the O_2 transfer reaction, indicating that the interfacial gas transfer process can significantly enhance mass transfer.

Studies on heterogeneous catalytic reactions typically focus on processes occurring at the interface between two phases: the solid catalyst and the reactant which may be in the gas or liquid state. In hydrocarbon oxidation reactions, where oxide lattice ions act as catalysts, additional phenomena such as the migration of oxygen ions and the filling of lattice vacancies in the presence of atmospheric oxygen are observed. In electrocatalytic reactions, charge migration occurs within the reactor in the form of ions, electrons, or protons, and the presence of the electrolyte introduces a third phase to the system.

6. Conclusions

Electrocatalytic processes, initially developed within the field of electrochemistry and not originally associated with catalytic reactions, have emerged as a distinct branch of catalysis. Interest in this area has grown significantly in recent years, driven by its applications in energy production and environmental protection. This chapter presents a comparative analysis of heterogeneous and electrochemical catalytic reactions, exploring the synergistic effects observed in the electrocatalytic systems, as well as their similarities and differences. The development and diversity of nanomaterials are closely linked to electrocatalytic processes, particularly in the context of fuel cells and their current and emerging applications and have been crucial in enhancing electrocatalytic activity, stability, and selectivity. Key factors such as electrode microstructure, three-phase boundary (TPB) density, and heterostructure design determine the efficiency of electrocatalytic transformations. Understanding these relationships provides both mechanistic insight and a theoretical framework for rational design of advanced electrocatalysts. Future research is expected to focus on optimizing the interplay between structural, electronic, and morphological properties of materials, integrating computational modeling with experimental studies, and developing sustainable electrocatalytic systems with improved performance and broader applicability in energy conversion and environmental remediation.

Conflict of interest

The authors declare no conflict of interest.

Author details

Viorica Parvulescu* and Gabriela Petcu
Institute of Physical Chemistry-Ilie Murgulescu, Romanian Academy, Bucharest, Romania

*Address all correspondence to: vpirvulescu@icf.ro

References

[1] Vayenas CG, Brosda S, Pliangos C. The double-layer approach to promotion, electrocatalysis, electrochemical promotion, and metal-support interactions. Journal of Catalysis. 2003;**213**:487-504. DOI: 10.1016/S0021-9517(02)00127-6

[2] Zhou C, Tao L, Yang F, Wang B, Wan X, Jin Y, et al. Application of electrochemical methods in heterogeneous catalysis. Current Opinion in Chemical Engineering. 2019;**26**:88-95. DOI: 10.1016/j.coche.2019.09.007

[3] Yamamoto K, Kuriyama M, Onomura O. Asymmetric electrosynthesis: Recent advances in catalytic transformations. Current Opinion in Electrochemistry. 2021;**28**:100714. DOI: 10.1016/j.coelec.2021.100714

[4] Zeng Y, Liang J, Zhong W, Li G, Deng D, Fan X, et al. Surface reconstruction regulation of catalysts for cathodic catalytic electrosynthesis. Applied Catalysis O. 2025;**202**:207036. DOI: 10.1016/j.apcato.2025.207036

[5] Wieckowski A, Neurock M. Contrast and synergy between electrocatalysis and heterogeneous catalysis. Advances in Physical Chemistry. 2011;**2011**:907129. DOI: 10.1155/2011/907129

[6] Fang Y-H, Liu Z-P. Tafel kinetics of electrocatalytic reactions: From experiment to first principles. ACS Catalysis. 2014;**4**:4364-4376. DOI: 10.1021/cs501312v

[7] Yin Q, Xu Z, Lian T, Musaev DG, Hill CL, Geletii YV. Tafel slope analyses for homogeneous catalytic reactions. Catalysts. 2021;**11**:87. DOI: 10.3390/catal11010087

[8] Huang J, Chen Y. Combining theory and experiment in advancing fundamental electrocatalysis. Current Opinion in Electrochemistry. 2019;**14**:A4-A9. DOI: 10.1016/j.coelec.2019.03.009

[9] Kori DKK, Das AK. Heterogeneous electrocatalysts for electrochemical cross-coupling reactions. ChemCatChem. 2024;**16**:202401491. DOI: 10.1002/cctc.202401491

[10] Seh ZW, Kibsgaard J, Dickens CF, Chorkendorff I, Nørskov JK, Jaramillo TF. Combining theory and experiment in electrocatalysis: Insights into materials design. Science. 2017;**355**(6321):4998. DOI: 10.1126/science.aad4998

[11] Vayenas CG. Electrochemical promotion of catalysis. Annals of Advanced Chemical Science. 2022;**6**:48-50. DOI: 10.29328/journal.aac.1001031

[12] Tsiplakides D, Balomenou S, Katsaounis A, Archonta D, Koutsodontis C, Vayenas CG. Electrochemical promotion of catalysis: Mechanistic investigations and monolithic electropromoted reactors. Catalysis Today. 2005;**100**:133-144. DOI: 10.1016/j.cattod.2004.12.015

[13] Yang S, Huang F, Yang H, Zhang J. The synergistic effects of cation vacancies in CoSe/NiSeheterojunctions and ternary CoNiSe catalysts for enhanced electrocatalytic and photoelectrochemical activities. Journal of Environmental Chemical Engineering. 2025;**13**:116691. DOI: 10.1016/j.jece.2025.116691

[14] Sheng Y, Yang R, Shi K, Yu H, Deng K, Wang Z, et al. Synergism of electrostatic attraction and tandem catalytic effect enabled efficient

electrosynthesis of ammonia from a wide-range of nitrate concentrations. Chemical Engineering Journal. 2024;**485**:149769. DOI: 10.1016/j.cej.2024.149769

[15] Wu X, Yan Q, Wang H, Wu D, Zhou H, Li H, et al. Heterostructured catalytic materials as advanced electrocatalysts: Classification, synthesis, characterization, and application. Advanced Functional Materials. 2024;**35**:2404535. DOI: 10.1002/adfm.202404535

[16] Neyman KM, Illas F. Theoretical aspects of heterogeneous catalysis: Applications of density functional methods. Catalysis Today. 2005;**105**:2-16. DOI: 10.1016/j.cattod.2005.04.006

[17] Han Y, Xu H, Li Q, Du A, Yan X. DFT-assisted low-dimensional carbon-based electrocatalysts design and mechanism study: A review. Frontiers in Chemistry. 2023;**11**:1286257. DOI: 10.3389/fchem.2023.1286257

[18] George J, Thangarasu S, Jayaram A, JesurajPeriyanayagam J. Materials for electrocatalysis: Future prospects in energy conversion. The Chemical Record. 2025;**25**:202400254. DOI: 10.1002/tcr.202400254

[19] Zhao M, Li M, Ding K, Bai Y, Liang X, Chen Y, et al. Thermal treatment of TiO_2–CuI mixtures to improve the electrochemical properties of TiO_2 anodes for lithium-ion batteries. International Journal of Electrochemical Science. 2025;**20**:101016. DOI: 10.1016/j.ijoes.2025.101016

[20] AshegaSherly R, Padma CM, Henry Raja D, Sindhusha S, Almansour AI, Dhas SJ, et al. H_2O_2-assisted photo-electrocatalytic and photocatalytic degradation of methyl violet by CuO-modified TiO_2 nanotube arrays. Optical Materials. 2024;**155**:115870. DOI: 10.1016/j.optmat.2024.115870

[21] Li J, Wang J, Wang G, Zhao S, Yang Z, Wang X, et al. CeO_2/ZnO heterojunction as efficient catalyst for electrocatalytic nitrogen reduction reaction via an “electron pump” effect. Journal of Rare Earths. DOI: 10.1016/j.jre.2025.03.012 [in press]

[22] Jin YX, Zhang XY, Zhang RX, Wang WJ, Zhao JX, Liu L, et al. Design and catalytic performance of Cu/γ-Al_2O_3 for electrocatalytic methanol oxidation reaction by using density functional theory and molecular dynamics. Applied Surface Science. 2024;**649**:159129. DOI: 10.1016/j.apsusc.2023.159129

[23] Alhashmialameer D, Shariq M, Qamar MA, Althikrallah HA, Al-Qasmi N, Al-abdali FH, et al. Improved catalytic performance of (Fe, Cr)-ZnO/g-C_3N_4 nanocomposite towards electrocatalytic water splitting for clean energy. Solid State Sciences. 2025;**160**:107823. DOI: 10.1016/j.solidstatesciences.2025.107823

[24] Mandari KK, Im Y, Bathula B, Lee YA, Kang M. Enhanced electrocatalytic oxygen evolution by silver-nickel-copper alloy nanoparticles supported on g-C_3N_4: A synergistic approach for improved catalytic performance. Journal of Alloys and Compounds. 2025;**1025**:180336. DOI: 10.1016/j.jallcom.2025.180336

[25] Kumaravel S, Avula B, Chandrasatheesh C, Niyitanga T, Saranya R, Hasan I, et al. Rational construction of MOF-derived α-Fe_2O_3/g-C_3N_4 composite for effective photocatalytic degradation of organic pollutants and electrocatalytic oxygen evolution reaction. SpectrochimicaActa Part A: Molecular and Biomolecular Spectroscopy. 2024;**310**:123972. DOI: 10.1016/j.saa.2024.123972

[26] Song Y, Han C, Li W, Yi X, Liao Q, Zhou J, et al. Engineering bimetallic cluster architectures: Harnessing unique "remote synergy effect" between Mn and Y for enhanced electrocatalytic oxygen reduction reaction. eScience. 2025;**5**:100332. DOI: 10.1016/j.esci.2024.100332

[27] Vijayaprabhakaran A, Sebastian C, Saravanakumar V, Vigneshwar I, Balaji R, Kathiresan M, et al. Cu-MOF and its multi-walled carbon nanotube composite for the effective catalytic and electrocatalytic hydrogenation of 4-nitrophenol. Journal of Alloys and Compounds. 2025;**1028**:180645. DOI: 10.1016/j.jallcom.2025.180645

[28] Li X, Huang J, Chu D, Cao L, Zhang X, Chen Q, et al. Catalytic C/Fe sites activation of Fe_3C embedded in carbon nanotubes by the incorporation of VN for multifunctional electrocatalytic properties. Energy. 2025;**319**:135045. DOI: 10.1016/j.energy.2025.135045

[29] Mao LX, Fu L, Zhang XY, Yan ZF, Gao ZH, Liu L, et al. Effect of transition metals (Ni and Co) in the methanol oxidation reaction electrocatalytic activity on Mo_2C/MWCNT. Molecular Catalysis. 2024;**567**:114455. DOI: 10.1016/j.mcat.2024.114455

[30] Ebrahimi P, Gholivand MB. Design of NiS@MnS/MWCNT heterostructure and its application for electrocatalytic oxidation of methanol with synergistically enhancing catalytic performance. International Journal of Hydrogen Energy. 2025;**118**:1-13. DOI: 10.1016/j.ijhydene.2025.02.162

[31] Wu Q, Cheng J, Sun Q, Yang Y, Yang S. Silver nanoparticles decorated ultrathin CoMo LDH nanosheets as hierarchical self-supported electrodes with high corrosion resistance for efficient seawater electrolysis. Journal of Alloys and Compounds. 2025;**1029**:180682. DOI: 10.1016/j.jallcom.2025.180682

[32] Song XZ, Ni JC, Wang XB, Dong JH, Liang HJ, Pan Y, et al. Hollow starlike Ag/CoMo-LDH heterojunction with a tunable d-band center for boosting oxygen evolution reaction electrocatalysis. Inorganic Chemistry. 2023;**62**:13328-13337. DOI: 10.1021/acs.inorgchem.3c01628

[33] Wei P, Zhang W, Luo M, Yu M, Zheng X, Duan R. Correlating crystallinity regulation in spinel oxides and catalytic activity in electrocatalytic water oxidation. Journal of Electroanalytical Chemistry. 2024;**971**:118565. DOI: 10.1016/j.jelechem.2024.118565

[34] Redondo J, Subbian J, Monclús MA, Pendashteh A, Pérez D, Mehdi M, et al. Effect of elastic strains on the electrocatalytic activity of Au thin films for the hydrogen evolution reaction. Journal of Catalysis. 2025;**447**:116151. DOI: 10.1016/j.jcat.2025.116151

[35] Heydari-Bafrooei E, Sabet M, Alizadeh Z, Ghazanfarpour R, Moazzen MH, Rezaei M, et al. Effect of Al and Zn impurities on the electrocatalytic activity of Nb2O5 toward hydrogen evolution reaction. Materials Chemistry and Physics. 2025;**332**:130254. DOI: 10.1016/j.matchemphys.2024.130254

[36] Chen L, Shen D, Li B, Xiao Z, Sun W, Liu X, et al. The efficient electrocatalytic and photocatalytic reduction of nitric oxide into ammonia over 0D/3D g-C3N4 quantum dots/3DOMM-TiO2-x heterojunction. Ceramics International. 2023;**49**:23129-23139. DOI: 10.1016/j.ceramint.2023.04.141

[37] Xu G, Zeng J, Tao J-L, Yao Y-F, Zhang H, Si L-P, et al. The effect of nitro group on the electrocatalytic hydrogen

evolution of triarylcorrole tin (IV) complex. Inorganica ChimicaActa. 2025;**585**:122749. DOI: 10.1016/j.ica.2025.122749

[38] Ma T, Qiu L, Tao Y, Cai Y, Liang F. Ligand dominating the catalytic performance of cucurbit[6]uril modified gold nanoparticles for electrocatalytic CO_2 reduction. Journal of Alloys and Compounds. 2024;**979**:173477. DOI: 10.1016/j.jallcom.2024.173477

[39] Sun H, Gao Y, Chen M, Li M, Xia Q, Liu Y, et al. Synergistic boosting the electrooxidation of biomass-based 5-hydroxymethylfurfural on cellulose-derived Co_3O_4/N-doped carbon catalysts. Materials Today Catalysis. 2024;7:100062. DOI: 10.1016/j.mtcata.2024.100062

[40] Wang L, Dou Y, Sun L, Huo L, Zhao H. Structural effects on 2e ORR electrocatalytic activity of LnFeO3 perovskites. Journal of Electroanalytical Chemistry. 2025;**987**:119113. DOI: 10.1016/j.jelechem.2025.119113

[41] Alhalili Z, Shariq M, Al-Qasmi N, Hakami O, Alathlawi HJ, Alharbi AF, et al. Enhanced catalytic activity of NiSe2 by nanohybrid formation with CoO nanosheets towards overall electrocatalytic water splitting for clean energy. International Journal of Hydrogen Energy. 2024;**94**:997-1004. DOI: 10.1016/j.ijhydene.2024.11.147

[42] Pappacena A, Bardini L, Boaro M. Catalysts and processes in solid oxide fuel cells. In: Boaro M, Salvatore AA, editors. Advances in Medium and High Temperature Solid Oxide Fuel Cell Technology. 1st ed. Switzerland: Springer; 2017. DOI: 10.1007/978-3-319-46146-5_5

[43] Feng R, Yang X, Ji W, Au C-T. Hydrothermal synthesis of stable mesoporousZrO_2-Y_2O_3 and CeO_2-ZrO_2-Y_2O_3 from simple inorganic salts and CTAB template in aqueous medium. Materials Chemistry and Physics. 2008;**107**:132-136. DOI: 10.1016/j.matchemphys2007.06.055

[44] Mather GC, Figueiredo FM, Jurado JR, Frade JR. Synthesis and characterisation of cermet anodes for SOFCs with a proton-conducting ceramic phase. Solid State Ionics. 2003;**162-163**:115-120. DOI: 10.1016/S0167-2738(03)00250-9

[45] Somacescu S, Calderon Moreno JM, Osiceanu P, Su B-L, Parvulescu V. Single-phase solid solution $(TiO_2)_x$-$(YSZ)_{1-x}$ mesoporous nanoparticles with catalytic activity in the oxidation of methane. Journal of Physical Chemistry C. 2010;**114**:19365-19372. DOI: 10.1021/jp105834a

[46] Cioatera N, Parvulescu V, Su B-L. Effect of nonionic surfactants on structure, texture and morphology of titania-doped YSZ. Materials Chemistry and Physics. 2010;**120**:697-701. DOI: 10.1016/j.matchemphys.2009.12.021

[47] Jiang SP, Chan SH. Development of Ni/Y_2O_3–ZrO_2 cermet anodes for solid oxide fuel cells. Materials Science and Technology. 2004;**20**:1109-1118. DOI: 10.1179/026708304225019957

[48] Gorte RJ, Vohs JM, McIntosh S. Recent developments on anodes for direct fuel utilization in SOFC. Solid State Ionics. 2004;**175**:1-6. DOI: 10.1016/j.ssi.2004.09.036

[49] Dong Y, Liu Z, Pang L, Han Y, Yao S, Wang X. Preparation of Y_2O_3–ZrO_2–CeO_2 solid solution by co-precipitation and its electrical property. Physica B: Condensed Matter. 2021;**612**:412972. DOI: 10.1016/j.physb.2021.412972

[50] Sasaki K, Takahashi I, Kuramoto K, Shin-mura K. A high-performance Ni-CeO_2/Ni/Ni-Y_2O_3•ZrO_2 three-layer anode for direct isooctane feeding of solid oxide fuel cells. Royal Society Open Science. 2022;**9**:220227. DOI: 10.1098/rsos.220227

[51] Somacescu S, Parvulescu V, Osiceanu P, Calderon-Moreno JM, Su B-L. Structure and surface chemistry in crystalline mesoporous (CeO2-$_\delta$)-YSZ. Journal of Colloid and Interface Science. 2011;**363**:165-174. DOI: 10.1016/j.jcis.2011.06.051

[52] Boshoman SB, Fatoba OS. Transition metal oxides as electrocatalytic material in fuel cells: A review. Journal of Engineering Science. 2023;**25**:948. DOI: 10.30919/es948

[53] Díaz-Aburto I, Hidalgo J, Fuentes-Mendoza E, González-Poggini S, Estay H, Colet-Lagrille M. Mo,Cu-doped CeO2 as anode material of solid oxide fuel cells (SOFCs) using syngas as fuel. Journal of Electrochemical Science and Technology. 2021;**12**(2):246-256. DOI: 10.33961/jecst.2020.01571

[54] Futamura S, Muramoto A, Tachikawa Y, Matsuda J, Lyth SM, Shiratori Y, et al. SOFC anodes impregnated with noble metal catalyst nanoparticles for high fuel utilization. International Journal of Hydrogen Energy. 2019;**44**:8502-8518. DOI: 10.1016/j.ijhydene.2019.01.223

[55] Somacescu S, Parvulescu V, Calderon-Moreno JM, Suh S-H, Osiceanu P, Su B-L. Uniform nanoparticles building Ce1-xPrxO2-$_\delta$ mesoarchitectures: Structure, morphology, surface chemistry and catalytic performance. Journal of Nanoparticle Research. 2012;**14**:1-17. DOI: 10.1007/s11051-012-0885-6

[56] Razpotnik T, Macek J. Synthesis of nickel oxide/zirconia powders via a modified Pechini method. Journal of the European Ceramic Society. 2006;**27**:1405-1410. DOI: 10.1016/j.jeurceramsoc.2006.05.059

[57] Cioateră N, Pârvulescu V, Rolle A, Vannier RN. Effect of strontium addition on europium-doped ceria properties. Solid State Ionics. 2009;**180**:681-687. DOI: 10.1016/j.ssi.2009.02.025

[58] Huan D, Shi N, Xie Y, Li X, Wang W, Xue S, et al. Cathode materials for proton-conducting solid oxide fuel cells. In: Kaur G, editor. Intermediate Temperature Solid Oxide Fuel Cells. Netherlands: Elsevier; 2020. pp. 263-314. DOI: 10.1016/B978-0-12-817445-6.00008-9

[59] Hossain S, Abdalla AM, Jamain SNB, Zaini JH, Azad AK. A review on proton conducting electrolytes for clean energy and intermediate temperature-solid oxide fuel cells. Renewable and Sustainable Energy Reviews. 2017;**79**:750-764. DOI: 10.1016/j.rser.2017.05.147

[60] Yu M, Feng Q, Liu Z, Zhang P, Zhu X, Mu S. Recent novel fabrication techniques for proton-conducting solid oxide fuel cells. Crystals. 2024;**14**:225. DOI: 10.3390/cryst14030225

[61] Sunarso J, Hashim SS, Zhu N, Zhou W. Perovskite oxides applications in high temperature oxygen separation, solid oxide fuel cell and membrane reactor: A review. Progress in Energy and Combustion Science. 2017;**61**:57-77. DOI: 10.1016/j.pecs.2017.03.003

[62] Zvonareva IA, Medvedev DA. Proton-conducting barium stannate for high-temperature purposes: A brief review. Journal of the European Ceramic Society. 2023;**43**:198-207. DOI: 10.1016/j.jeurceramsoc.2022.10.049

[63] Tang Z, Xie Y, Hawthorne H, Ghosh D. Sol-gel processing of Sr0.5Sm0.5CoO3 film. Journal of Power Sources. 2006;**157**:385-388. DOI: 10.1016/j.jpowsour.2005.07.041

[64] Lei Y, Fang J, Xu Y, Wang X, Xia C, Wang B, et al. Low-temperature direct ethanol solid oxide fuel cells based on LiNi0.8Co0.15Al0.05O2-$_{\delta}$ electrodes. International Journal of Hydrogen Energy. 2025;**130**:242-248. DOI: 10.1016/j.ijhydene.2025.04.284

[65] Halim EM, Chemchoub S, El Attar A, Salih FE, Oularbi L, El Rhazi M. Recent advances in anode metallic catalysts supported on conducting polymer-based materials for direct alcohol fuel cells. Frontiers in Energy Research. 2022;**10**:843736. DOI: 10.3389/fenrg.2022.843736

[66] Wang Y, Zhou D, Zhang W, Yang J, Sheng Q, Zhu X, et al. CoO_x composite $Pr_{0.4}Sr_{0.6}Co_{0.2}Fe_{0.8}O_{3-\delta}$ perovskite was used to obtain an IT-SOFC cathode with high electrocatalytic activity. Fuel. 2025;**381**:133422. DOI: 10.1016/j.fuel.2024.133422

[67] Donazzi A, Schmauss TA, Barnett SA. Catalytic and electrocatalytic performance of $Sr(Ti_{0.3}Fe_{0.7}Ru_{0.07})O_{3-\delta}$ for applications in solid oxide fuel cells supplied with ethanol steam reforming mixtures. Journal of Power Sources. 2022;**551**:232215. DOI: 10.1016/j.jpowsour.2022.232215

[68] Yang J, Zhang Y, Zhu X, Bai J, Li S, Yan W, et al. Introducing transition metal cations to promote the oxygen reduction kinetics of double perovskite as SOFC cathode. Chemical Engineering Journal. 2025;**516**:164058. DOI: 10.1016/j.cej.2025.164058

[69] Gao Y, Zhang M, Fu M, Hu W, Tong H, Zetian TZ. A comprehensive review of recent progresses in cathode materials for proton-conducting SOFCs. Energy Reviews. 2023;**2**:100038. DOI: 10.1016/j.enrev.2023.100038

[70] Gómez SY, Hotza D. Current developments in reversible solid oxide fuel cells. Renewable and Sustainable Energy Reviews. 2016;**61**:155-174. DOI: 10.1016/j.rser.2016.03.005

[71] Lu Y, Noor A, Ahmed J, Alwadie N, Akhtar MN, Abid S, et al. Synergistic effects and electrocatalytic insight of single-phase hexagonal structure as low-temperature solid oxide fuel cell cathode. Journal of Rare Earths. 2025;**43**:1390-1399. DOI: 10.1016/j.jre.2024.06.027

[72] Ramadhani F, Hussain MA, Mokhlis H, Hajimolana S. Optimization strategies for solid oxide fuel cell (SOFC) application: A literature survey. Renewable and Sustainable Energy Reviews. 2017;**76**:460-484. DOI: 10.1016/j.rser.2017.03.052

[73] Wang X, Cheng SA, Feng YJ, Merrill MD, Saito T, Logan BE. Use of carbon mesh anodes and the effect of different pretreatment methods on power production in microbial fuel cells. Environmental Science & Technology. 2009;**43**:6870-6874. DOI: 10.1021/es900997w

[74] Qian F, Wang GM, Li Y. Solar-driven microbial photoelectrochemical cells with a nanowire photocathode. Nano Letters. 2010;**10**:4686-4691. DOI: 10.1021/nl102977n

[75] Rezaei A, Sisi AJ, Zarenezhad H, Aber S, Khataee A. Photo/electrocatalytic microbial fuel cell for simultaneous pollutant degradation and electricity production using $g\text{-}C_3N_4$ anchored on Ag-mIm. Process Safety and Environmental Protection. 2023;**180**:1132-1144. DOI: 10.1016/j.psep.2023.10.009

[76] Antoniadou M, Kondarides D I, Labou D, Neophytides S, Lianos P. An efficient photoelectrochemical cell functioning in the presence of organic wastes. Solar Energy Materials & Solar Cells. 2010;**94**:592-597. DOI: 10.1016/j.solmat.2009.12.004

[77] He Y, Chen K, Leung MKH, Zhang Y, Li L, Li G, et al. Photocatalytic fuel cell – A review. Chemical Engineering Journal. 2022;**428**:131074. DOI: 10.1016/j.cej.2021.131074

[78] Zhao S, Xu S. Semiconductor photoanode photoelectric properties of methanol fuel cells. Journal of Nanoelectronics and Optoelectronics. 2021;**16**(1):72-79. DOI: 10.1166/jno.2021.2906

[79] Mureseanu M, Chivu V, Osiac M, Ciobanu M, Bucur C, Parvulescu V, et al. New photoactive mesoporousCe-modified TiO2 for simultaneous wastewater treatment and electric power generation. Catalysis Today. 2021;**366**:164-176. DOI: 10.1016/j.cattod.2020.09.035

[80] Li K, Yang X, Guo YN, Ma FY, Li HC, Chen L, et al. Design of mesostructured $H_3PW_{12}O_{40}$-titania materials with controllable structural orderings and pore geometries and their simulated sunlight photocatalytic activity towards diethyl phthalate degradation. Applied Catalysis B: Environmental. 2010;**99**:364-375. DOI: 10.1016/j.apcatb.2010.07.012

[81] Meena SL, Meena K, Bhupesh RK, Meena RC. Novel study on 2-D material in photoelectrochemical cell for the conversion and storage of solar energy. Oriental Journal of Chemistry. 2023;**39**:627-634. DOI: 10.13005/ojc/390313

[82] Sànchez de Bustamante L, Aguadero A, Fernàndez-Díaz MT, Santos Silva Jr S, Biškup N, Martínez JL, et al. Dual doping synergy: Optimizing $SrMoO_3$ perovskite anodes via in-situ Ni exsolution and Cr doping for enhanced SOFC efficiency. Journal of Alloys and Compounds. 2025;**1018**:179116. DOI: 10.1016/j.jallcom.2025.179116

[83] Huang Y, Feng J-H, Yi Y, Duan F, Liu X-H, Tian G-Y, et al. Recent progress of direct ammonia fuel cells: From materials to transportation applications. Rare Metals. 2025;**44**(6):3686-3708. DOI: 10.1007/s12598-024-03126-5

[84] Stoeckl B, Preininger M, Subotić V, Schroettner H, Sommersacher P, Seidl M, et al. Ammonia as promising fuel for solid oxide fuel cells: Experimental analysis and performance evaluation. ECS Transactions. 2019;**91**:1601-1610. DOI: 10.1149/09101.1601ecst

[85] Schüth F, Palkovits R, Schlögl R, Su DS. Ammonia as a possible element in an energy infrastructure: Catalysts for ammonia decomposition. Energy & Environmental Science. 2012;**5**:6278-6289. DOI: 10.1039/c2ee02865d

[86] Zheng K, Yan Y, Sun Y, Yang J, Zhu M, Ni M, et al. An experimental study of ammonia decomposition rates over cheap metal catalysts for solid oxide fuel cell anode. International Journal of Hydrogen Energy. 2023;**48**:19188-19195. DOI: 10.1016/j.ijhydene.2023.01.312

[87] Lei S, Wang A, Weng G, Wu Y, Xue J, Wang H. Simultaneous generation of electricity, ethylene and decomposition of nitrous oxide via protonic ceramic fuel cell membrane reactor. Journal of Energy Chemistry. 2023;**77**:359-368. DOI 10.1016/j.jechem.2022.10.035

[88] Millet P. Electrochemical membrane reactors, current trends and future developments on (bio) membranes. In: Basile A, Millet P, editors. Recent Advances on Membrane

Reactors. Netherlands: Elsevier; 2023. pp. 285-313. DOI: 10.1016/B978-0-12-823659-8.00009-5

[89] Stoukies M. Solid-electrolyte membrane reactors: Current experience and future outlook. Catalysis Reviews Science and Engineering. 2000;**42**:1-70. DOI: 10.1081/CR-100100259

[90] Gorte RJ, Vohs JM. Novel SOFC anodes for the direct electrochemical oxidation of hydrocarbons. Journal of Catalysis. 2003;**216**:477. DOI: repository.upenn.edu/handle/20.500.14332/5898

[91] Cruz-Martínez H, Rojas-Chávez H, Matadamas-Ortiz PT, Ortiz-Herrera JC, López-Chávez E, Solorza Feria O, et al. Current progress of Pt-based ORR electrocatalysts for PEMFCs: An integrated view combining theory and experiment. Materials Today Physics. 2021;**19**:100406. DOI: 10.1016/j.mtphys.2021.100406

[92] Qian W, Xu S, Zhang X, Li C, Yang W, Bowen CR, et al. Differences and similarities of photocatalysis and electrocatalysis in two-dimensional nanomaterials: Strategies, traps, applications and challenges. Nano-Micro Letters. 2021;**13**:156. DOI: 10.1007/s40820-021-00681-9

[93] Kong W, Zhang M, Han Z, Zhang Q. A theoretical model for the triple phase boundary of solid oxide fuel cell electrospun electrodes. Applied Sciences. 2019;**9**:493. DOI: 10.3390/app9030493

[94] Li Y, Wu Z, Wang C, Yu X, Gao W, Wang B, et al. Engineering triple-phase boundary in Pt catalyst layers for proton exchange membrane fuel cells. Advanced Functional Materials. 2024;**34**:2310428. DOI: 10.1002/adfm.202310428

[95] Ryu CH, Mandal D, Ren H. Gas-liquid-solid three-phase boundary in scanning electrochemical cell microscopy. ACS Measurement Science Au. 2024;**4**:729-736. DOI: 10.1021/acsmeasuresciau.4c00061

[96] Wang Y-C, Huang W, Wan L-Y, Yang J, Xie R-J, Zheng Y-P, et al. Identification of the active triple-phase boundary of a non-Pt catalyst layer in fuel cells. Science Advances. 2022;**8**:eadd8873. DOI: 10.1126/sciadv.add8873

[97] Shaukat S, Tabish AN, Irshad M, Akbar S, Farhat I, Fan L. A comprehensive study on effective triple-phase boundary density and its correlation with active anode thickness in solid oxide fuel cells. Energy Science & Engineering. 2024;**12**:3883-3894. DOI: 10.1002/ese3.1850

Chapter 2

Function of Catalyst-Material Phases in Low-Temperature Electrolysis

Ilias Stamatelos and Worawee Saei

Abstract

Low-temperature electrolysis driven by renewable energy sources has emerged as a vital route for sustainable chemical production, carbon cycle closure, and the realization of a circular economy. Central to the advancement of this technology is the development of efficient and stable electrocatalyst materials, which govern both the thermodynamics and kinetics of electrochemical reactions. The catalytic phase—specifically whether the material exists as a metal, oxide, or hydroxide—is decisive in determining key performance metrics such as activity, selectivity, stability, and product distribution. Each phase exhibits unique physicochemical attributes, including variations in crystal structure, electronic properties, and surface affinity, affecting adsorption behavior, charge transfer dynamics, and corrosion resistance. This chapter provides a detailed exploration of the roles of metal, oxide, and hydroxide phases in electrocatalysis, focusing on their mechanistic contributions in critical reactions like the electrochemical CO_2 reduction (ECR) and alkaline oxygen evolution reaction (OER). In addition to discussing the intrinsic properties of each phase, the chapter highlights the significance of phase evolution, particularly in situ transformations and induced transitions, as a promising strategy to enhance catalytic performance. These dynamic processes often lead to increased surface area and grain boundary density, which can substantially boost reactivity and durability. By combining structural insights with mechanistic understanding, this chapter aims to offer a cohesive framework for designing next-generation electrocatalysts optimized for high-performance electrolysis applications.

Keywords: electrochemistry, CO_2 reduction, low-temperature electrolysis, electrocatalyst, oxygen evolution reaction, OER, alkaline electrolysis

1. Introduction

Low-temperature electrolysis, particularly when powered by renewable energy sources, is rapidly emerging as a cornerstone technology in the transition to a circular and decarbonized economy. By facilitating the transformation of plentiful and readily accessible feedstocks—including water and captured carbon dioxide (CO_2)—into chemical fuels and valuable industrial products, it provides a direct pathway to reduce greenhouse gas emissions, enhance energy security, and close the anthropogenic carbon cycle. The process is underpinned by two electrochemical reactions: the electrochemical reduction of carbon dioxide (ECR) at the cathode and the oxygen evolution reaction (OER) at the anode. Collectively, these elements facilitate the generation of storable and transportable energy carriers, including hydrogen and carbon-based fuels, thereby addressing the mounting demand for flexible, distributed energy systems [1–3].

In recent years, ECR has attracted significant attention as a method of storing excess renewable electricity in the form of chemical bonds. The process has the potential to yield a range of valuable products, including carbon monoxide (CO), formate, methane, ethanol, and other C2+ compounds, with the particular product depending strongly on the catalyst used. Among these, CO and formate are regarded as particularly promising due to their relatively high selectivity and favorable techno-economic profiles. More complex products, such as ethanol or hydrocarbons, offer high energy density and are therefore desirable for liquid fuel applications. However, their large-scale production remains challenging due to competing reaction pathways and low Faradaic efficiencies [4–7]. In parallel, water electrolysis continues to serve as the most mature and direct route to green hydrogen production. In this process, the cathodic half-reaction, otherwise known as the hydrogen evolution reaction (HER), is relatively facile, while the anodic half-reaction, otherwise known as the OER, remains kinetically sluggish due to its complex four-electron mechanism [8]. Consequently, the OER is widely regarded as the limiting step in overall water-splitting efficiency and is a key focus of catalyst development efforts. The advancement of both ECR and OER is contingent on the design and implementation of effective electrocatalyst materials [9–11]. The catalyst's properties determine the energy required to drive the reaction and also influence reaction kinetics, intermediate stabilization, product selectivity, and long-term system durability. With regard to the reduction of CO_2, the composition of the catalyst dictates the favored pathways. For example, copper uniquely facilitates C-C coupling and the formation of C_2 + products, while silver and bismuth favor simpler products such as CO and formate [12]. In the context of the OER, traditional benchmarks such as IrO_2 and RuO_2 offer excellent performance in acidic and alkaline environments. However, they are constrained by scarcity and cost, rendering their large-scale application unsustainable. This has given rise to a significant increase in the level of interest in transition metal oxides and hydroxides, particularly those based on nickel, iron, and cobalt, which have demonstrated competitive activity and are more abundant. However, these alternatives frequently exhibit limited durability under prolonged operation, undergoing structural transformations or dissolution under oxidative electrochemical conditions [13–16].

The development of electrocatalysts capable of achieving high activity, selectivity, and stability simultaneously remains a central challenge in electrolysis research. Achieving product specificity in ECR is a particularly complex process due to the numerous competing multi-electron pathways, each of which is sensitive to changes in catalyst morphology, surface structure, and local reaction environment. It is imperative to consider stability, particularly in the context of OER catalysts, which are required to demonstrate resilience to harsh oxidative potentials, fluctuating pH levels, and corrosive electrolytes over the course of their operational lifetimes. A considerable number of materials that show great promise in laboratory settings often demonstrate significant degradation when subjected to industrial conditions. Moreover, the overall system efficiency is closely associated with the overpotential required to initiate and sustain the reactions. It is imperative to reduce these energy inputs while preserving elevated current densities to ensure economic viability and facilitate scale-up [17–21].

In order to address these interconnected challenges, it is necessary to develop a more profound mechanistic understanding of the behavior of catalysts under operating conditions. This includes phase transitions, active site restructuring, and interfacial dynamics. As electrochemical systems move toward practical deployment, these efforts will be instrumental in bridging the gap between fundamental materials science and applied electrochemical engineering. It is therefore vital to acknowledge the potential of low-temperature electrolysis to play a pivotal role in the development of a resilient, low-carbon energy infrastructure.

2. Fundamentals of electrocatalysis in electrolysis

The ECR is a complex, multistage process that encompasses the transfer of multiple electrons [1–3]. It involves several crucial steps, including CO_2 adsorption, electron transfer, and product desorption occurring at the electrode's surface. Extensive research has revealed that the current density, the specific electrochemical species involved, and the selectivity of the ECR are predominantly influenced by the choice of electrode material and the reduction potential applied [22–24]. The initial step of the ECR is the diffusion of $CO_{2(g)}$ over the catalyst layer, followed by one e-transfer for the formation of the $\bullet CO^{-}_{2(ads)}$ intermediate, Eq. (1). The following protonation of the $\bullet CO^{-}_{2(ads)}$ Eq. (2), aided by the water molecule, is proposed to happen rapidly, forming the transition •COOH species [25]. A subsequent step involves both an electron and a proton transfer, Eq. (3), forming the adsorbed $\bullet CO_{(ads)}$. Finally, the $\bullet CO_{(ads)}$ desorbs from the catalyst's surface, yielding the $CO_{(g)}$ Eq. (4) [26–28].

$$\mathrm{CO_{2(g)}} + \bullet + \mathrm{e}^- \leftrightarrow \bullet CO^{-}_{2(ads)} \tag{1}$$

$$\bullet CO^{-}_{2(ads)} + \mathrm{H_2O_{(l)}} \leftrightarrow \bullet \mathrm{COOH_{(ads)}} + \mathrm{OH}^- \tag{2}$$

$$\bullet \mathrm{COOH_{(ads)}} + \mathrm{H}^+ + \mathrm{e}^- \leftrightarrow \bullet \mathrm{CO_{(ads)}} + \mathrm{H_2O_{(l)}} \tag{3}$$

$$\bullet \mathrm{CO_{(ads)}} \leftrightarrow \mathrm{CO_{(g)}} + \bullet \tag{4}$$

The electrochemical CO_2 reduction (ECR) process follows the Eley-Rideal mechanism, where only one species undergoes adsorption [29, 30]. Initial Tafel analysis data suggested that Eqs. (1) and (2) could be the possible rate-determining steps (RDS) for ECR. However, recent work by Deng et al. [31] has shown that the RDS for CO formation is equation Eq. (1), while the RDS is independent of the catalyst material used and the reaction environment. The Eq. (1) is a composite step that encompasses the crucial processes of $CO_{2(g)}$ diffusion over the catalyst layer and electron transfer. At lower current densities, the rate of electron transfer in Eq. (1) is the critical factor influencing the rate of ECR. At higher levels of polarization, of more than 200 mA cm^{-2}, the diffusion of $CO_{2(g)}$, as seen in Eq. (1), toward the active surface of the catalyst becomes the limiting factor for the ECR rate. Given the strong correlation between the catalyst's properties and the ECR performance, the Sabatier principle emerges as a crucial tool for predicting a catalyst's activity and product selectivity [32]. It asserts that for high catalytic activity, the interactions between the catalyst and the reaction species should exhibit an intermediate intensity for the reaction to be favored. The parasitic hydrogen evolution reaction (HER) progresses at the cathode in parallel with the ECR. The generally accepted HER mechanism in alkaline and neutral media is exhibited below:

$$H_2O + e^- \leftrightarrow H_{ads} + OH^-, \text{(Volmer step)} \quad (5)$$

$$H_{ads} + H_2O + e^- \leftrightarrow H_2 + OH^-, \text{(Heyrovsky step)} \quad (6)$$

$$2H_{ads} \leftrightarrow H_2, \text{(Tafel step)} \quad (7)$$

The splitting of a water molecule and the adsorption of hydrogen on the surface of the electrode or catalyst is described by the Volmer step, Eq. (5), which always consists of the initial step of the HER. This is then followed by the production of hydrogen through either an electrochemical process, the Heyrovsky step, Eq. (6), or a chemical process, as in a Tafel step, Eq. (7). Numerous studies have exhibited that at low η, lower than −0.15 V, the HER goes through the Heyrovsky step, while the Tafel step follows in parallel. However, at a high η of more than −0.25 V, the electrode is considered overwhelmed by H_{ads}. The Tafel step becomes negligible, with the HER progressing through the Heyrovsky step, owing to its high rate constant [33–35]. The low thermodynamic potential of the HER, summarized in **Table 1**, and its faster overall kinetics strongly favor the reaction at the expense of ECR (**Figure 1**).

At lower polarization, the reactions' kinetics are mainly determined by the e-transfer. The function of a selective and efficient catalyst for the ECR is to successfully redirect the majority of the total current of the electrolyzer toward the ECR instead of the HER. It is worth noting that at a high current density of more than 300 mA cm^{-2}, both the ECR, Eq. (4), and HER, Eq. (6), rates are primarily governed by species desorption, with additional limitations imposed by mass transfer of the reactants. Considering the significantly higher diffusion coefficient of H_2, D_{H2}, compared to the

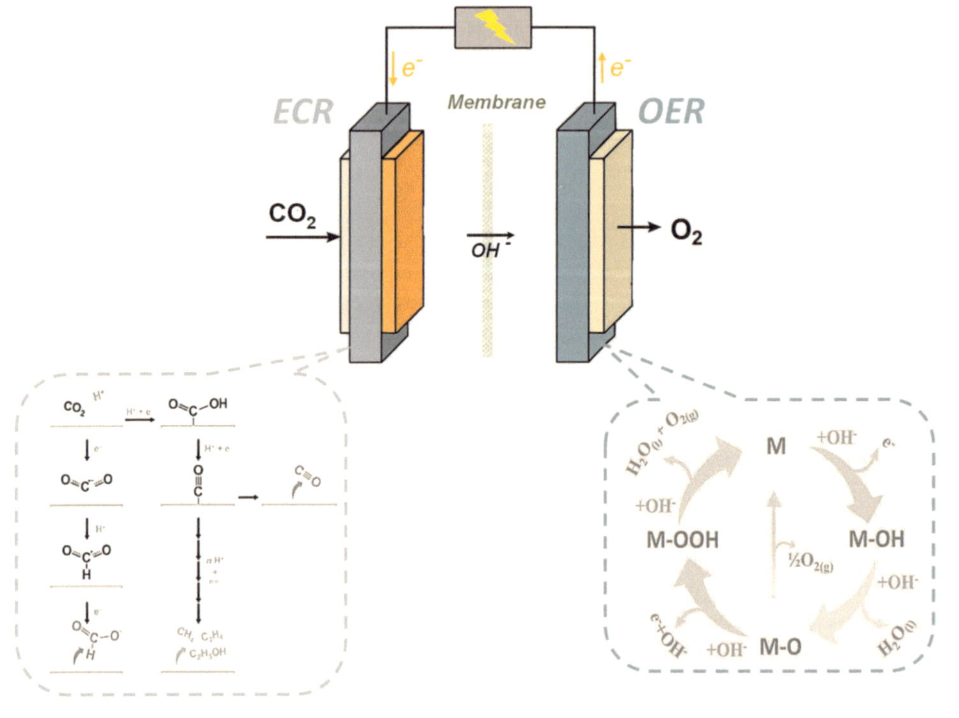

Figure 1.
Schematic representation of a typical electrolyzer, utilized for the cathodic CO_2 valorization (ECR) and the anodic oxygen production (OER) at alkaline conditions.

Product (Reaction)	Equation	E_o (V vs. SHE, 25°C)
Hydrogen (HER, acidic)	$2H^+ + 2e^- \rightarrow H_2$	0
Carbon monoxide (ECR)	$CO_2 + 2H^+ + 2e^- \rightarrow CO + H_2O$	−0.104
Formic acid (ECR)	$CO_2 + 2H^+ + 2e^- \rightarrow HCOOH$	−0.171
Methanol (ECR)	$CO_2 + 6H^+ + 6e^- \rightarrow CH_3OH + H_2O$	0.016
Methane (ECR)	$CO_2 + 8H^+ + 8e^- \rightarrow CH_4 + 2H_2O$	0.169
Ethylene (ECR)	$2CO_2 + 12H^+ + 12e^- \rightarrow C_2H_4 + 4H_2O$	0.085
Oxygen (OER, alkaline)	$4OH^- \rightarrow O_2 + 2H_2O + 4e^-$	0.401
Hydrogen (HER, alkaline)	$2H_2O + 2e^- \rightarrow H_2 + 2OH^-$	−0.828

Table 1.
Half-reaction potentials of the main low-temperature electrolysis paths [36].

CO_2 one, D_{CO2}, and CO one, D_{CO} [37], the faster mass transfer of the H_2 will inevitably favor the HER at high current densities. At lower reaction rates, ECR is favored by the catalyst's active centers so that the partial current density for the ECR, j_{CO}, is always larger than that for HER, j_{H2}. While at higher currents, due to the effect of mass-transfer limitations on the reactions' species, HER takes over, imposing the opposite condition [38–40]. Three key figures of merit determine the preference for either ECR or HER. The Faradaic Efficiency, FE $(\%) = \frac{Qpr}{Qth}$, quantifies the ratio of the theoretically calculated charge spent on product formation Q_{th}, calculated by the total charge passed, to the actual amount Q_{pr}, quantified by analytical techniques. The partial current density j_i signifies the rate of product formation ($j_i = j$*FE), derived from the total current density flowing through the electrolyzer j and the FE for the specific product (i) formation. Evaluating the viability of the reaction, the energy efficiency, EE $(\%) = \frac{Eo}{Ecell}$ FE of the electrolyzer for a specific product (i), offers valuable insights. It articulates the thermodynamic efficiency of product formation, as expressed through the relation $\frac{Eo}{Ecell}$, relative to its kinetic efficiency (FE).

The oxygen evolution reaction (OER) in alkaline media, found in **Table 1**, follows complex yet distinct mechanistic pathways as reaction pathways [41, 42]. The two primary mechanistic pathways have been identified for the OER in alkaline media as either the Adsorbate Evolution Mechanism (AEM) or the Lattice Oxygen Mechanism (LOM). The AEM is the traditional surface-mediated pathway where OER proceeds through proton-coupled electron transfers (PCETs) involving adsorbed intermediates [2, 43]. These steps typically occur on transition metal active sites (M), with oxygen forming via a sequence of reactions involving *OH, *O, and *OOH surface species, as seen below:

$$OH^- + M \rightarrow M\text{–}OH + e^-, (OH^- \text{ adsorption and oxidation}) \quad (8)$$

$$M\text{–}OH + OH^- \rightarrow M\text{–}O + H_2O + e^-, (\text{Formation of oxo intermediate}) \quad (9)$$

$$M\text{–}O + OH^- \rightarrow M\text{–}OOH + e^-, (O - O \text{ bond formation, RDS}) \quad (10)$$

$$M\text{–}OOH + OH^- \rightarrow O_2 + H_2O + M + e^-, (O_2 \text{ release and regeneration of active site}) \quad (11)$$

The AEM activity is closely tied to the binding energies of these intermediates. However, the adsorption energies of *OH and *OOH are linearly correlated, with a typical energy difference of ~3.2 eV. This imposes a theoretical overpotential limit of ~370 mV. To overcome this bottleneck, recent strategies focus on optimizing intermediate adsorption via doping, vacancy engineering (V_O), strain modulation, and electronic structure tuning, according to the Sabatier principle [44]. In contrast, the LOM introduces an alternative pathway in which the lattice oxygen (O_{lat}) of the catalyst itself participates in the OER. This mechanism circumvents the *OOH intermediate and breaks the linear scaling constraints of AEM, potentially enabling lower overpotential and faster kinetics, as seen:

$$M\text{–}O_{lat} \rightarrow M + V_O + \bullet O_{lat} \text{ (Formation of surface oxygen radicals)} \quad (12)$$

$$OH^- + M \rightarrow M\text{–}OH + e^- \quad (13)$$

$$M\text{–}OH + \bullet O_{lat} \rightarrow O_2 + M + H^+ + e^- \text{ (Coupling of adsorbed OH with } O_{lat}\text{)} \quad (14)$$

$$\text{or}: 2O_{lat}^- \rightarrow O_2 + 2e^- \text{ (Alternatively, } O_{lat} - O_{lat} \text{ coupling)} \quad (15)$$

$$V_O + OH^- \rightarrow O_{lat} + e^- \text{ (} O_{la} \text{ position regeneration)} \quad (16)$$

While LOM offers the advantage of enhanced activity by overcoming the adsorption energy limitations of AEM, it may compromise catalyst stability due to irreversible lattice oxygen loss and the formation of oxygen vacancies. Nevertheless, it has been shown that tuning metal-oxygen co-valence can modulate lattice participation, with lower Tafel slopes (~43 mV dec^{-1}) serving as an experimental fingerprint for LOM pathways [45, 46].

Overall, it is evident that the catalyst material plays a pivotal role in shaping both the reaction pathway and the overall efficiency of the electrochemical process, with its properties directly influencing performance and selectivity.

3. Electrocatalytic phases

Catalyst-material phases—namely metals, oxides, and hydroxides—play crucial and often complementary roles in governing electrocatalytic performance in low-temperature electrolysis, especially for the ECR and the OER. These phases influence key factors such as electron conductivity, adsorption energetics, intermediate stabilization, and structural stability, which in turn define activity, selectivity, and durability [47].

3.1 Metal phases in electrocatalysis

Metallic phases serve as highly conductive active sites, promoting efficient electron transfer and the adsorption of reactants. Their electrocatalytic performance is fundamentally influenced by their intrinsic properties, including electronic structure, binding energies, and the position of the d-band center. The electronic structure dictates the distribution of electronic states in the vicinity of the Fermi level, which in turn affects the ability of the catalyst to adsorb and activate reactants. The binding energies of the key intermediates must be optimized to ensure their stability during the reaction while also allowing for the desorption of the product. A pivotal descriptor in this context is the d-band center, whose position relative to the Fermi level is indicative of the adsorption strength of reaction intermediates. Metals with a higher d-band center have been observed to bind adsorbates more strongly, which has the potential to enhance activation. However, this binding may also impede product release. These properties have been shown to influence the overall activity, selectivity, and stability of metal catalysts [48].

The selectivity of ECR is governed by the adsorption energies and reaction pathways of key intermediates on the catalyst surface. For instance, the formation of *COOH and *OCHO intermediates leads to CO and formate products, respectively. The binding strength of these intermediates is influenced by the electronic structure of the catalyst, particularly the d-band center. A higher d-band center is generally associated with enhanced adsorption, which can stabilize intermediates but concurrently impede product desorption. Conversely, a lower d-band center leads to weaker adsorption, facilitating desorption but potentially destabilizing intermediates. It is notable that copper (Cu) is distinguished by its capacity to yield multicarbon (C_2+) products, including ethylene and ethanol. This phenomenon is attributed to its moderate binding strength for *CO intermediates, which facilitates C-C coupling steps. The formation of C-C bonds is a critical step in the generation of C2+ products, and it requires a delicate balance in the adsorption energies of intermediates to proceed efficiently. Conversely, silver (Ag) and gold (Au) exhibit high selectivity toward CO production due to their weaker binding of *CO intermediates, which promotes desorption before further reduction can occur. Tin (Sn) has been observed to promote the formation of formate, a process that is believed to be the result of its strong affinity for the *OCHO intermediate. The aforementioned trends in product selectivity are closely tied to the metal-dependent adsorption energetics of intermediates such as *COOH, *CO, and *OCHO [49]. A key addition from Ringe et al. [50] is the recognition of the role played by electrochemical double-layer (EDL) effects and field-driven adsorption in ECR kinetics, particularly for metals like gold. Their multiscale modeling and experimental analysis demonstrated that the rate-limiting step for CO formation on Au shifts with overpotential, from *COOH to *CO at low overpotential to CO_2 adsorption at intermediate potentials, and eventually to CO_2 mass transport at high overpotential. This underscores the significance of EDL-induced surface charge effects on *CO_2 stabilization and the recognition that reaction kinetics are not solely dictated by thermodynamics or static binding energies but also by field-mediated interactions with dipolar intermediates. In a similar manner, the alloying of metals has been demonstrated to modulate the electronic structure and binding properties of the active surface, thereby exerting a consequential influence on ECR selectivity. For instance, Cu-Ag alloys have been demonstrated to enhance C2 product formation by stabilizing *CO and *OCCO intermediates while suppressing formate pathways. This synergistic effect is attributable to the electronic interactions between Cu and Ag atoms, which adjust the d-band center and, consequently, the adsorption energies of key intermediates. In the same context, Cu-Sn alloys have been observed to exhibit selectivity toward CO or formate production, depending on the composition. The incorporation of Sn into Cu modifies the electronic environment, affecting the binding strength of intermediates and altering reaction pathways. It is possible to design catalysts with tailored selectivity for desired ECR products by carefully controlling the alloy composition and structure [51, 52].

Transition metal materials in their metallic phases (e.g., Ni, Co, Fe, Cu, or their alloys) are typically characterized by high electrical conductivity and structural robustness. These properties enhance OER efficiency by lowering energy barriers and accelerating intermediate transformations (e.g., M–OH → M–O → M–OOH) [53, 54]. However, as OER catalysts, their direct activity is generally limited. Metallic phases frequently function as precursors or conductive supports, forming the backbone structure that undergoes in situ transformation into more catalytically surface-active oxide or oxy-hydroxide phases under OER operating conditions. This process maintains mechanical stability and prevents degradation, as evidenced by the NiFe/NiFeOx core-shell structure [55]. Transition metals dynamically shift oxidation states (e.g., $Ni^0 \rightarrow Ni^{3+}$, $Fe^0 \rightarrow Fe^{3+}$), optimizing adsorption energies of OER intermediates (*O, *OH, *OOH) and fine-tuning reaction pathways [56]. The combination of metals with differing oxidation states, such as NiFe alloys with Fe^{3+} and Ni^0, has been shown to yield synergistic effects, thereby reducing the overpotential to 160 mV at 10 mA/cm^2. This is achieved by leveraging Fe^{3+} for OER activity and Ni^0 for charge transfer. In the context of OER, metallic surfaces oxidize to form active oxyhydroxides (e.g., NiOOH, CoOOH), with the metallic core ensuring the continuous supply of electrons to these catalytic sites [57]. Strategies such as doping (e.g., Fe into Ni) or hybridization with conductive supports further enhance performance by modifying electronic structures, stabilizing intermediates, and increasing surface area. Collectively, these mechanisms ensure the balance of conductivity, active site availability, and stability, thus enabling transition metal-based catalysts to demonstrate comparable OER efficiency to that of noble metal systems [58].

3.2 Oxide phases in electrocatalysis

Oxide phases, typically transition metal oxides (e.g., NiO, Co_3O_4, Cu_2O), offer tunable redox chemistry and rich surface chemistry. Their electrocatalytic functionality is governed by a combination of their electronic conductivity and surface reactivity, both of which are highly dependent on their composition and structural features. Metal oxides exhibit a wide spectrum of electronic conductivities, ranging from insulating to semiconducting to metallic, rendering them versatile platforms for catalytic applications. Transition metal oxides exhibit high intrinsic conductivity, facilitating efficient charge transfer during reactions such as the ECR or the OER [59]. The surface reactivity of metal oxides is influenced by surface terminations, lattice defects, and the presence of hydroxyl groups, complementing their bulk properties. The surface features of the material under investigation create active sites that enhance the adsorption and activation of reactant molecules. It is noteworthy that oxygen vacancies can function as electron-rich sites, thereby promoting CO_2 adsorption and facilitating the formation of key intermediates during reduction. Furthermore, the acid-base character of the oxide surface has been demonstrated to modulate the adsorption strength and stability of intermediates, thereby directing reaction pathways and influencing product selectivity. Collectively, these intrinsic and tunable properties render metal oxides as highly adaptable materials for electrocatalytic systems [60].

For the ECR, metal oxides fulfill a vital function by stabilizing key reaction intermediates, thereby significantly influencing both catalytic activity and product selectivity. In the context of reductive electrochemical conditions, metal oxides frequently undergo partial reduction, resulting in the generation of distinctive surface characteristics that diverge significantly from their metallic counterparts. These features include the formation of oxygen vacancies, the emergence of mixed-valence states, and the retention of metastable oxide phases. Each of these contributes to the catalytic behavior observed in oxide-containing electrocatalysts [61]. One of the most critical contributions of partially reduced metal oxides is the stabilization of key reaction intermediates through the presence of oxygen vacancies (VO). These electron-rich defects function as active adsorption sites, thereby facilitating the binding and activation of CO_2 molecules. The enhancement of the formation and stabilization of intermediates such as *COOH and *OCHO is influenced by oxygen vacancies, which in turn impact the subsequent formation of CO and formate. To illustrate this point, consider the case of SnO2-based catalysts. The deliberate generation of oxygen vacancies in these catalysts has been shown to correlate with an enhancement in formate selectivity. This is due to the stabilization of the *OCHO intermediate, which occurs preferentially over competing pathways [62]. Similarly, ZnO surfaces have been demonstrated to exhibit high CO selectivity, even in their partially reduced states. This finding suggests that active oxide sites persist under operational conditions and contribute directly to catalytic performance [63].

Furthermore, partially reduced metal oxides often exhibit a mixture of oxidation states on their surfaces, as exemplified by oxide-derived copper catalysts that feature Cu^0, Cu^+, and Cu^{2+} species [64]. This heterogeneity creates a dynamic and flexible surface environment where various intermediates can be stabilized through different adsorption configurations. The presence of Cu + sites has been shown to be specifically correlated with enhanced C-C coupling, a critical step in the formation of multicarbon products such as ethylene and ethanol. In addition to stabilizing intermediates, the surface characteristics of partially reduced oxides have been shown to suppress competing reactions, notably the HER. The utilization of catalysts that exhibit a preference for the adsorption and conversion of CO_2-derived species over protons has been demonstrated to enhance the Faradaic efficiency toward the formation of desired carbon-based products. The capacity of non- and partially reduced metal oxides to provide tailored active sites, facilitate multi-electron reaction pathways, and mitigate undesired side reactions positions them as highly effective and tunable materials for selective and efficient ECR.

In addition to pristine metal oxides, composite heterostructured oxide-based catalysts have demonstrated considerable application potential as efficient ECR catalysts. A variety of metal-oxide interfaces are of significance in this regard, not only as structural foundations but also as contributors to the overall catalytic performance [65]. At these junctions, electronic and structural synergies arise that modify the adsorption energies of CO_2 and its intermediates, enhance local electric fields, and tune the d-band center of the metallic phase. These interfacial phenomena create energetically favorable pathways for key reaction steps such as *COOH or *CO stabilization, C-C coupling, and proton-electron transfer [66]. Besides promoting direct catalytic activity through interfacial synergy, metal oxides serve a crucial function as catalyst supports for the ECR. Oxides like Cr_2O_3 and CeO_2 offer excellent chemical and electrochemical stability under the reductive potentials typically applied during CO_2 reduction. These materials are capable of maintaining their structural integrity and surface composition while enabling the uniform dispersion and anchoring of ECR active metallic nanoparticles (e.g., Ag) [67]. The stability of these supports under reaction conditions is essential for preventing nanoparticle aggregation and maintaining high electrochemical surface area throughout long-term operation.

Beyond their structural role, these oxides can also influence the local electronic environment at the metal-support interface. A characteristic example is the CeO_2, known to possess oxygen storage capacity and a reversible Ce^{4+}/Ce^{3+} redox couple, which enables it to participate in dynamic surface charge transfer processes during catalysis. This can influence the electronic density of deposited metal particles, leading to modified intermediate binding and potentially altered product selectivity [21]. These oxide-supported systems offer several strategic advantages: (i) improved dispersion of supported catalytic sites, (ii) enhanced CO_2 adsorption at oxide-metal boundaries, and (iii) the suppression of side reactions (e.g., HER) through surface polarity tuning. In addition, these supports often contribute to a more favorable electrochemical environment by affecting local ion distributions and double-layer capacitance, both of which are known to impact ECR kinetics [68].

Transition metal oxides (TMOs) have been identified as a class of materials with considerable potential for use as OER catalysts, particularly within an alkaline environment. The appeal of these materials can be attributed to three key factors. Firstly, their structural diversity plays a significant role in their appeal, as do their tunable electronic properties. Secondly, their ability to host a variety of oxidation states is a crucial factor. Transition metal oxides (TMOs) have the capacity to adopt a variety of crystal structures, including spinel (e.g., Co_3O_4), perovskite (e.g., $LaNiO_3$), and layered (e.g., NiO). These structures offer distinct catalytic sites and pathways for oxygen evolution. A significant development in the field of TMO-based OER catalysis is the emergence of mixed metal oxides, where the amalgamation of two or more transition metals results in a synergistic effect. For instance, NiFe oxides have repeatedly exhibited superior OER activity in comparison to their single-metal counterparts. The incorporation of Fe into NiO or Ni(OH)2 matrices has been demonstrated to increase the number of active sites while concomitantly modulating the electronic structure, thereby optimizing the adsorption energies of OER intermediates such as *OH, *O, and *OOH [69–71]. Similarly, the doping of other metals can introduce additional active sites, create oxygen vacancies, and enhance charge transfer. Cu-doped Co oxides have also demonstrated reduced overpotential and improved stability, attributed to lattice distortions and enhanced surface reactivity [72].

Defect engineering is another powerful strategy for boosting the OER performance of TMOs. Introducing oxygen vacancies, cation deficiencies, or lattice strain can significantly alter the electronic structure and increase the density of active sites [73]. For example, porous $NiO/NiCo_2O_4$ nanotubes with abundant oxygen vacancies have been shown to facilitate a four-electron OER pathway, achieving low Tafel slopes and high current densities [74]. The presence of defects not only enhances intrinsic activity but also improves conductivity, which is often a limiting factor for oxide catalysts. Morphology control, such as the synthesis of aerogels, nanowires, or hollow structures, further amplifies the catalytic performance by maximizing surface area and exposing more active facets. NiFe oxide aerogels, for instance, combine the benefits of high porosity and interconnected networks, leading to rapid mass transport and efficient charge separation [75]. Oxide phases offer a versatile platform for OER catalysis, with their performance tunable through compositional, structural, and morphological modifications. The ongoing development of mixed metal oxides, defect-rich materials, and hybrid structures continues to push the boundaries of OER efficiency and durability.

3.3 (Oxy)-hydroxide phases in electrocatalysis

Metal hydroxides represent a distinct and functionally versatile class of electrocatalytic materials, particularly in alkaline media, where they often serve as active phases or dynamic precursors to catalytically competent species. Their layered structures, rich in hydroxyl groups (-OH), enable reversible proton and electron transfer, thereby conferring remarkable electronic and chemical adaptability. These characteristics render them particularly pertinent in both the OER and the ECR contexts.

The emergence of hydroxide phases as critical components in ECR has been particularly pronounced in alkaline and near-neutral pH environments. The function of these elements extends beyond the provision of structural support, as they contribute directly to key aspects of the catalytic process, including intermediate stabilization, interfacial charge modulation, and promotion of multicarbon product pathways. These materials frequently manifest as surface layers or discrete domains in hybrid systems, exhibiting distinctive electrochemical and chemical properties that enhance both activity and selectivity. A pivotal mechanism through which hydroxide phases promote ECR is through the modification of the local pH near the catalyst surface. This effect is especially significant when hydroxide species are formed or accumulated in situ during electrochemical operation. As demonstrated in the research conducted by Xiao et al., the development of hydroxide-rich layers on Cd-based catalysts has been shown to significantly increase the local pH level. This, in turn, has been observed to impede the competing hydrogen evolution reaction (HER) and to encourage the formation of carbon monoxide (CO) with Faradaic efficiencies that exceed 99% across a wide range of potential ranges [76]. The catalytic enhancement can be mechanistically attributed to several factors. First, the elevated local pH favors the deprotonation and activation of CO_2 molecules, facilitating the formation of key intermediates such as *COOH and *CO. These intermediates benefit from hydrogen bonding interactions with surface hydroxyls, which stabilize their adsorption and lower the energy barrier for subsequent reaction steps. In particular, *CO stabilization is crucial for efficient CO desorption or further reduction into multicarbon products on metal centers such as Cu. Secondly, hydroxide phases actively participate in forming dynamic and adaptive surface structures. Under ECR conditions, many metal catalysts can develop surface-reconstructed layers consisting of mixed metal-oxide-hydroxide domains. These layers serve as electron-rich interfaces, which alter the local electric field and improve the adsorption energetics of intermediates. For instance, in Cu-based catalysts, the presence of hydroxide has been linked to enhanced C-C coupling due to increased *CO coverage and stabilization of *OCCO intermediates—critical steps in the formation of C_2 + products [77, 78]. Additionally, hydroxide-containing interfaces such as metal-hydroxide junctions have been shown to enhance charge redistribution and interfacial electric fields, tuning the binding energy of intermediates. These properties enable not only improved CO_2 activation but also selective pathway steering by inhibiting proton adsorption and HER, enhancing the ECR. The structural and chemical flexibility of hydroxides under operating conditions allows them to function as dynamic catalytic components. They can regenerate or reconfigure during operation in response to potential changes, thereby maintaining active sites and suppressing deactivation phenomena observed in other oxide or metallic systems [79–81].

While oxide phases have traditionally been regarded as the primary OER catalysts, recent *in situ* and in *operando* studies have revealed that, under OER operating conditions, many TMOs and metallic precursors undergo surface reconstruction to form oxy-hydroxide phases (e.g., NiOOH, CoOOH, FeOOH). These oxyhydroxides have emerged as the true active phases responsible for the highest OER activity, particularly in alkaline media. The formation of oxy-hydroxide phases is a dynamic process, often initiated by the application of anodic potentials during OER. For example, Ni $(OH)_2$ is oxidized to NiOOH, while $Co(OH)_2$ transforms into CoOOH [82]. The presence of Fe is especially beneficial; Fe-doped NiOOH (NiFeOOH) exhibits dramatically enhanced activity compared to pure NiOOH. Fe incorporation induces lattice strain, modulates the electronic structure, and optimizes the adsorption energies of OER intermediates, resulting in lower overpotential and higher turnover frequencies [83, 84]. Recent *in operando* X-ray absorption and Mössbauer spectroscopy studies have identified high-valent Fe^{4+} species as key active sites in NiFe oxyhydroxides, persisting even after prolonged OER operation [85, 86]. Introducing chromium (Cr) into the phase transition of metal (Fe, Co, or/and Ni) hydroxides creates an FeCoNiCr multimetal hydroxide phase to enhance the active OER phase, oxy-hydroxide, [87] by stabilizing high-valent Co^{3+}/Ni^{3+} states, according to recent research. This synergistic effect surpasses the enhancements observed in Fe-doped systems alone. The presence of Cr promotes the formation of a more active β-NiOOH phase, leading to improved OER kinetics. Furthermore, Cr can regulate the crystallization and phase homogenization of NiFe and CoFe oxide/hydroxides, inducing the formation of smaller, more homogenous nanodots [87–89].

Layered double hydroxides (LDHs) are recognized as highly effective pre-catalysts for the OER in alkaline water electrolyzers due to their unique structural and electronic properties. The layered structure of LDHs, composed of edge-sharing metal-hydroxide octahedra with intercalated anions, allows for significant compositional flexibility and the incorporation of multiple metal ions. This versatility enables LDHs to undergo dynamic surface reconstruction under OER conditions, transforming into catalytically active metal oxy-hydroxide phases such as NiOOH or FeOOH. To illustrate this point, consider the example of NiFe-LDHs, which can reconfigure into highly active NiFeOOH phases. This process results in low overpotential and enables expeditious redox cycling of the metal centers [90, 91]. Additionally, the high surface area of LDHs, especially when exfoliated into single-layer nanosheets, greatly increases the number of accessible active sites, leading to significantly enhanced OER activity compared to bulk materials and even surpassing benchmark catalysts like IrO_2 [92, 93]. The compositional flexibility of LDHs also allows for synergistic interactions between different metal ions, such as Ni and Fe in NiFe-LDHs, where Ni promotes charge transfer and Fe optimizes the adsorption energies of OER intermediates. This synergy results in catalytic performance that is outstanding, with LDHs demonstrating low overpotential at high current densities. In addition, the hierarchical and porous architectures frequently observed in LDH-based materials enhance mass transport and stability, thereby ensuring efficient operation even at industrially relevant current densities. It is evident that the aforementioned characteristics collectively render LDHs a versatile and powerful platform for the development of next-generation OER catalysts in water electrolyzers [94].

Oxy-hydroxides are characterized by higher electrical conductivity than their oxide counterparts, owing to the presence of hydrated layers and proton intercalation. This facilitates rapid transport and efficient electron transfer during OER. The electronic and structural properties of oxyhydroxides can be further tuned through doping, defect engineering, and the creation of heterostructures. For example, Mo-doped NiFeOOH has been reported to follow a heterogeneous diatomic oxygen mechanism, achieving exceptional activity in alkaline media [95–97]. While oxyhydroxides are highly active in alkaline environments, their stability in acidic media remains a challenge due to the dissolution of transition metal ions. Overall, oxy-hydroxide phases are now widely recognized as the true active catalysts for OER, with their formation and stability being key determinants of long-term performance. The dynamic interplay between oxide and oxy-hydroxide phases underscores the importance of *in situ* and *in operando* characterization techniques in understanding and optimizing OER catalysts.

4. Phase transitions and in situ catalyst evolution

Electrocatalytic systems frequently undergo dynamic phase transitions (**Figure 2a**) between metallic, oxide, and hydroxide states, thereby significantly

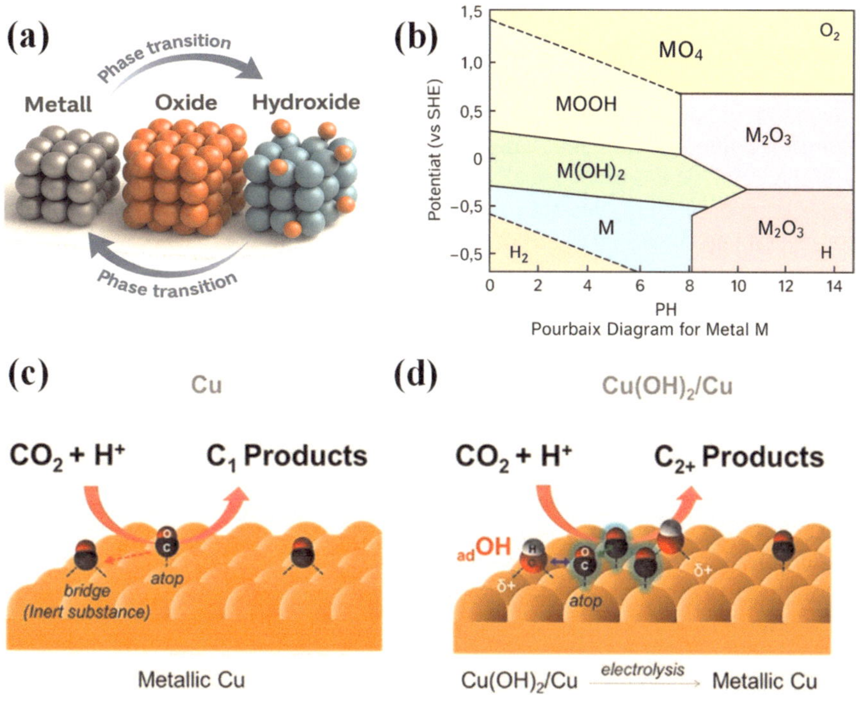

Figure 2.
(a) Schematic representation of the main phase transitions to be observed in redox reactions for the different material phases. (b) Generic Pourbaix diagram for a metal (M) indicating the various phase transitions induced by overpotential and pH. (c) ECR reaction pathway over metallic Cu. (d) ECR reaction pathway over metallic-Cu(OH)$_2$ and its parallel non-Faradaic reduction to metallic-Cu; both figures (c and d) were reprinted after permission from Iijima et al. [98].

influencing catalytic activity, selectivity, and stability. The electrochemical environment is a driving force behind these transformations and is fundamental to key processes such as water splitting, nitrogen reduction, and carbon conversion. The Pourbaix diagram (**Figure 2b**) provides a thermodynamic framework for predicting the stable phase of a material as a function of pH and electrode potential.

4.1 Phase transitions in CO_2 reduction electrocatalysis

Phase transitions among metal, oxide, and hydroxide states play a crucial role in governing catalytic behavior during the ECR. These dynamic transformations, strongly influenced by parameters such as applied potential, electrolyte composition, and pH, can significantly alter the catalyst's surface composition, electronic structure, and reactivity [99]. One of the most impactful transitions is the reduction of metal oxides and hydroxides to their metallic state (**Figure 2c** and **d**) under cathodic conditions. This process gives rise to oxide-derived metals (OD-Ms), a class of catalysts that exhibit distinct physicochemical properties compared to their pristine metallic counterparts. The formation typically begins with the electrochemical or thermal reduction of an oxide-based material, acting as the pre-catalyst [100]. This principle is applied to both pristine metal-oxide materials and composite oxide pre-catalysts. Leading to the formation of either single-phase OD-Ms or OD-metal alloys [101, 102]. This redox cycling results in the partial or complete removal of lattice oxygen, generating defects such as vacancies, dislocations, and grain boundaries. These structural features enhance the density of active sites and alter the local electronic environment, while residual subsurface oxygen or oxide species may persist and further influence catalytic behavior [103, 104]. The transformation is often incomplete due to kinetic limitations, particularly in thicker oxide layers or under mild reduction conditions. The result is a mixed-phase surface composed of metallic, oxide, and/or hydroxide domains. These interfacial regions create a heterogeneous and highly active catalytic landscape, enabling alternative reaction pathways and offering a broad spectrum of binding sites not present on clean metal surfaces. Oxide-derived metals possess several key characteristics that make them particularly effective for ECR. Their defect-rich morphology, high surface area, and high grain boundary (GB) concentration improve CO_2 adsorption and intermediate stabilization, and overall, the ECR activity [105, 106]. The modified electronic structure—often reflected in shifts in the d-band center due to embedded oxygen species—alters the binding energies of crucial intermediates such as *COOH, *CO, and *OCHO, thus tuning both activity and product selectivity [107]. Furthermore, metal-oxide interfaces present in these materials contribute to intermediate stabilization via local electric fields and spatial confinement. Importantly, the partially reduced and defective surfaces also suppress the competing HER, thereby enhancing Faradaic efficiency for carbon-based products.

In addition to oxide-to-metal transitions, the reversible transformation between metal oxides and hydroxides plays an equally crucial role in tuning catalytic behavior for the ECR. The electrolyte pH has a significant influence on the stability and restructuring of these phases [81]. Under alkaline conditions, metal oxides readily convert to hydroxides due to the abundance of OH^- ions, whereas acidic environments favor re-oxidation or dehydroxylation back to oxide phases. These phase transitions significantly affect the surface speciation and electronic properties of the catalyst, directly impacting the adsorption energies of key intermediates and, consequently, the activity and selectivity of the ECR. Notably, the formation of hydroxide layers during ECR is not a passive or inert phenomenon but a dynamic, in situ process that creates highly reactive surface states [98]. Hydroxide-rich layers have been shown to elevate the local pH, suppress the competing HER, and stabilize critical intermediates such as *CO and *OCCO—key precursors in C-C coupling pathways. These effects collectively promote the formation of valuable multicarbon (C_2^+) products such as ethylene and ethanol, especially under alkaline electrolysis conditions. Hydroxide-derived metals (HD-Ms), obtained via electrochemical reduction of pre-formed metal hydroxides, exhibit catalytic properties distinct from their pristine metallic counterparts [108]. The reduction process introduces a high density of surface defects, such as vacancies, grain boundaries, and amorphous regions, that serve as catalytic hot spots. These structural features facilitate CO_2 adsorption and activation, while residual hydroxyl or subsurface oxygen species can modulate the electronic structure, finely tuning the binding strength of reaction intermediates. This is particularly evident in copper-based systems, where surface-adsorbed hydroxyls (OH_{ad}) can either enhance *CO adsorption and C-C coupling when optimally present or hinder activity by blocking active sites and promoting HER if in excess [76]. The hydroxide-to-metal transformation, therefore, is not merely a preparative step but a functional transition that defines the catalyst's performance. The structural and electronic complexity resulting from this transition enables enhanced selectivity toward C_2^+ products by facilitating multistep reaction pathways and intermediate stabilization [109].

4.2 Phase transitions in oxygen evolution reaction (OER)

The OER in water electrolysis requires efficient electrocatalysts to overcome its inherently sluggish kinetics. LDHs have emerged as promising pre-catalysts due to their ability to undergo in situ structural transformation into active metal oxy-hydroxide phases under OER conditions. LDHs, with their general formula $[M^{2+}{}_{1-x}M^{3+}{}_{x}(OH)_2][A^{n-}]_{x/n}\cdot mH_2O$, feature tunable metal cations (e.g., Ni, Fe, Co) and interlayer anions (e.g., NO_3^-, CO_3^{2-}) that enable dynamic reconstruction during electrochemical activation. Under anodic potentials, LDHs transition from their as-synthesized α-phase to catalytically active γ-phase oxyhydroxides (MOOH), characterized by ~8% lattice contraction and interlayer cation exchange (e.g., K^+ replacing anions). This structural evolution enhances electrical conductivity and exposes redox-active metal sites, as observed in NiFe-LDHs that form γ-NiFeOOH with optimized Ni^{3+}/Fe^{3+} coordination environments [110].

The phase transformation is driven by oxidative deprotonation and metal oxidation, facilitated by defects and dopants. For instance, Cr doping in FeCoNi hydroxides reduces the free energy barrier for $M(OH)_2 \rightarrow MOOH$ conversion by 10–22%, enabling faster activation [111]. Similarly, oxygen vacancies in $Sr_{.2}$–$Ca_xFe_2O_6$–δ perovskites boost conductivity by two orders of magnitude compared to defect-free analogs. *In situ* electrochemical atomic force microscopy reveals that strain-induced cracking and wrinkling in NiCo-LDH nanosheets during OER correlate with partial oxidation to $NiCoO(OH)_x$, highlighting the interplay between structural dynamics and activity [112]. These reconstructed oxyhydroxides follow a Mars-van Krevelen mechanism, where lattice oxygen participates in O-O bond formation via O-bridged Fe-M reaction centers (M = Ni, Co), lowering the energy barrier for *OOH deprotonation [113].

The effectiveness of LDHs as pre-catalysts originates from their layered structure, which ensures uniform distribution of metal cations and provides efficient ionic diffusion pathways. Dynamic interlayer anions such as NO_3^- disrupt the layered framework during electrochemical activation, generating structural disorder that produces highly defective interfaces. These defect-rich surfaces exhibit optimized electronic configurations and coordinative unsaturated sites, which directly boost catalytic performance by lowering activation barriers for intermediate adsorption [114]. Post-transformation, the resulting oxyhydroxides exhibit exceptional stability (e.g., FeCoNiCr hydroxide maintains 150-hour OER activity with a 224 mV overpotential) [111]. This *in situ* activation mechanism highlights the adaptability of LDHs as customizable frameworks for advancing high-efficiency OER catalysts.

The dynamic transformation between metal oxides and oxyhydroxides plays a key role in determining the performance and stability of OER electrocatalysts. Under anodic potentials, oxides such as NiO and Co_3O_4 transform into active oxy-hydroxide phases (e.g., γ-NiOOH, CoOOH), which possess hydrated layered structures that facilitate efficient proton-coupled electron transfer and expose more active sites [115]. The stabilization of these oxy-hydroxide species can be enhanced through doping and strain engineering; for example, Fe incorporation in NiOOH improves redox cycling and durability, while interstitial Si doping in RuO_2 suppresses unwanted lattice oxygen activity and enhances stability in acidic media [116]. Cation leaching and surface restructuring also play crucial roles. Controlled leaching, such as Mo from NiFe sulfides, can generate defect-rich, highly active surfaces and extend catalyst lifetime, but excessive leaching leads to structural degradation and loss of activity. Therefore, balancing reconstruction and leaching is key to achieving both high activity and long-term stability in OER catalysts. These findings underscore the importance of tuning oxide-hydroxide transformations and surface chemistry to design robust, efficient water-splitting electrocatalysts. These degradation pathways are exacerbated in acidic media, where lattice oxygen participation in OER (Mars-van Krevelen mechanism) induces crystalline phase collapse and irreversible surface amorphization [117].

4.3 Comparative analysis of phase interplay and stability

In the ECR, the stability and performance of catalysts are intricately linked to phase transformations occurring during operation, particularly in oxide-derived and hydroxide-transformed metal systems. Oxide-derived metals, such as copper, undergo partial reduction (e.g., $Cu_2O \rightarrow Cu^0$) under cathodic conditions, creating highly defective surfaces enriched with grain boundaries and undercoordinated sites that enhance activity and C-C coupling efficiency. However, these features also introduce vulnerabilities, such as structural coarsening, surface atom migration, and residual oxide instability, which compromise long-term durability under sustained operation [118–120]. Phase transitions between hydroxides and metallic copper (e.g., $Cu(OH)_2 \rightarrow Cu^0$) also yield catalytically favorable morphologies but can suffer from dynamic reconstruction and uncontrolled subsurface oxidation or re-oxidation, especially at fluctuating potentials. Additionally, electrolyte-driven restructuring, including carbonate formation or cation exchange (e.g., K^+ or Na^+ adsorption), can lead to loss of active phase, catalyst delamination, and suppression of multicarbon product pathways [121, 122].

To address these degradation pathways, recent strategies have focused on controlling redox transformations and interfacial composition. Pre-conditioning oxide catalysts under reducing potentials enables the formation of thermodynamically stabilized defect sites while minimizing over-reduction or collapse of active regions [123]. Doping approaches have been shown to modulate electronic structures and maintain surface roughness essential for C-C coupling while simultaneously mitigating agglomeration and phase collapse. Composite catalyst systems, particularly metal-oxide and metal-hydroxide hybrids (e.g., $Cu\text{-}CeO_2$, $Cu\text{-}Ni(OH)_2$), further enhance operational robustness by offering lattice strain buffering and dual-functionality in CO_2 activation and proton management. These heterojunctions facilitate charge redistribution and stabilize reactive intermediates like *CO or *CHO through synergistic interfacial fields. Operando studies reveal that maintaining partial oxide features under bias—rather than a complete reduction—is critical for preserving selectivity toward C_2 + products [124–126].

The interplay between oxide and oxy-hydroxide phases during OER directly impacts catalytic stability and degradation pathways. In operational conditions, catalysts are subject to dynamic structural transformations, including hydroxylation of metal oxides (e.g., $NiO \rightarrow \gamma\text{-}NiOOH$) and sulfur-oxygen exchange in sulfides. These processes expose active sites and concomitantly trigger destabilizing processes, such as cation leaching and phase over-oxidation [127].

Strategies to stabilize active phases focus on modulating reconstruction pathways. Doping high-valent cations (e.g., Mo in NiFeOOH) strengthens metal-oxygen bonds, suppressing over-oxidation and maintaining γ-phase integrity under high potentials [128]. Similarly, strain engineering in perovskites like $SrCo_{0.9}Fe_{0.1}O_3\text{-}\delta$ stabilizes metastable phases with optimal oxygen vacancy densities, enhancing conductivity while resisting segregation [129]. For layered double hydroxides (LDHs), Cr doping promotes homogeneous cation distribution, delaying phase separation during OER and enabling 150-hour stability [130]. These approaches highlight the balance required between activity-enhancing phase transitions and structural preservation, guiding the design of catalysts that reconcile high performance with operational durability [131, 132].

DOI: http://dx.doi.org/10.5772/intechopen.1011365

5. Conclusion

The ECR is highly sensitive to the structural and electronic properties of the catalyst, which are governed by dynamic phase transformations. Oxide-derived metals (O-DMs) have emerged as highly effective electrocatalysts due to their defect-rich surfaces and altered electronic structures resulting from oxide-to-metal reduction. These structural features stabilize key intermediates such as *CO and *OCCO, thereby promoting multicarbon (C_{2+}) product formation. Subsurface oxygen residues and grain boundaries in O-DMs have been shown to shift the d-band center, thereby tuning the adsorption energies of intermediates and enhancing reaction selectivity. Furthermore, in situ studies demonstrate that such catalysts are not static but undergo dynamic restructuring under operating conditions, thereby continuously regenerating active sites. Furthermore, the integration of oxides such as CeO_2 or ZnO within composites has been demonstrated to enhance CO_2 activation through the introduction of synergistic effects and oxygen vacancies, thereby facilitating enhanced electron transfer. While contemporary strategies have considerably enhanced selectivity and activity, challenges persist in the domains of long-term stability control and atomic-level phase behavior comprehension.

The catalytic performance of transition metal-based materials for the OER is intricately linked to their phase composition and dynamic transformations under operating conditions. Metallic phases function as conductive precursors, thereby facilitating the formation of catalytically active surfaces. Oxide phases offer structural versatility and tunable electronic properties, with their performance further enhanced by defect engineering, morphology control, and compositional tuning. In conclusion, it is the oxy-hydroxide phases that are formed in situ during OER that host the most active sites and drive the reaction with the highest efficiency, particularly when doped synergistically with elements such as Fe or Cr.

Recent advances in in situ and in operando characterization have enhanced the understanding of dynamic phase transitions in electrocatalysts, enabling more rational and targeted design strategies. Recent research has focused on the use of doping, defect engineering, and strain modulation to direct surface reconstruction and optimize active site exposure and electronic structure. In future research, the focus should be on enhancing the stability of oxy-hydroxide phases, particularly in acidic environments. Future efforts should also focus on developing hybrid materials with tunable interfaces, leveraging computational modeling, and applying machine learning to predict and control phase behavior. The integration of real-time diagnostics to monitor structural evolution under operating conditions has the potential to further accelerate the development of efficient, stable, and earth-abundant catalysts, thereby advancing the deployment of sustainable electrochemical energy technologies.

Author details

Ilias Stamatelos[1,2]* and Worawee Saei[1,2]

1 Forschungszentrum Jülich GmbH, Institute of Electrochemical Process Engineering (IET-4), Jülich, Germany

2 RWTH Aachen University, Aachen, Germany

*Address all correspondence to: i.stamatelos@fz-juelich.de

References

[1] Laha S, Lee Y, Podjaski F, Weber D, Duppel V, Schoop LM, et al. Ruthenium oxide nanosheets for enhanced oxygen evolution catalysis in acidic medium. Advanced Energy Materials. 2019;**9**(15): 1803795. DOI: 10.1002/aenm.201803795

[2] Grimaud A, Hong WT, Shao-Horn Y, Tarascon J-M. Anionic redox processes for electrochemical devices. Nature Materials. 2016;**15**(2):121-126. DOI: 10.1038/nmat4551

[3] Middleton RS, Carey JW, Currier RP, Hyman JD, Kang Q, Karra S, et al. Shale gas and non-aqueous fracturing fluids: Opportunities and challenges for supercritical CO_2. Applied Energy. 2015; **147**:500-509. DOI: 10.1016/j.apenergy.2015.03.023

[4] Spurgeon JM, Kumar B. A comparative technoeconomic analysis of pathways for commercial electrochemical CO_2 reduction to liquid products. Energy & Environmental Science. 2018;**11**(6):1536-1551. DOI: 10.1039/C8EE00097B

[5] Verma S, Kim B, Jhong H-R"M", Ma S, Kenis PJA. A gross-margin model for defining technoeconomic benchmarks in the electroreduction of CO_2. ChemSusChem. 2016;**9**(15):1972-1979. DOI: 10.1002/cssc.201600394

[6] Dinh C-T, García de Arquer FP, Sinton D, Sargent EH, Rate H. Selective, and stable electroreduction of CO_2 to CO in basic and neutral media. ACS Energy Letters. 2018;**3**(11):2835-2840. DOI: 10.1021/acsenergylett.8b01734

[7] Weng L-C, Bell AT, Weber AZ. A systematic analysis of Cu-based membrane-electrode assemblies for CO_2 reduction through Multiphysics simulation. Energy & Environmental Science. 2020;**13**(10):3592-3606. DOI: 10.1039/D0EE01604G

[8] King LA, Hubert MA, Capuano C, Manco J, Danilovic N, Valle E, et al. A non-precious metal hydrogen catalyst in a commercial polymer electrolyte membrane electrolyser. Nature Nanotechnology. 2019;**14**(11):1071-1074. DOI: 10.1038/s41565-019-0550-7

[9] García de Arquer FP, Dinh C-T, Ozden A, Wicks J, McCallum C, Kirmani AR, et al. CO_2 electrolysis to multicarbon products at activities greater than 1 a Cm^{-2}. Science. 2020;**367**(6478):661-666. DOI: 10.1126/science.aay4217

[10] Wang X, Wang Z, García de Arquer FP, Dinh C-T, Ozden A, Li YC, et al. Efficient electrically powered CO_2-to-ethanol via suppression of deoxygenation. Natural Energy. 2020;**5**(6):478-486. DOI: 10.1038/s41560-020-0607-8

[11] Zhong M, Tran K, Min Y, Wang C, Wang Z, Dinh C-T, et al. Accelerated discovery of CO_2 electrocatalysts using active machine learning. Nature. 2020; **581**(7807):178-183. DOI: 10.1038/s41586-020-2242-8

[12] Kortlever R, Shen J, Schouten KJP, Calle-Vallejo F, Koper MTM. Catalysts and reaction pathways for the electrochemical reduction of carbon dioxide. Journal of Physical Chemistry Letters. 2015;**6**(20):4073-4082. DOI: 10.1021/acs.jpclett.5b01559

[13] Trasatti S. Electrocatalysis by oxides — Attempt at a unifying approach. Journal of Electroanalytical Chemistry and Interfacial Electrochemistry. 1980; **111**(1):125-131. DOI: 10.1016/S0022-0728(80)80084-2

[14] Lee Y, Suntivich J, May KJ, Perry EE, Shao-Horn Y. Synthesis and activities of

rutile IrO_2 and RuO_2 nanoparticles for oxygen evolution in acid and alkaline solutions. Journal of Physical Chemistry Letters. 2012;**3**(3):399-404. DOI: 10.1021/jz2016507

[15] Son YJ, Kawashima K, Wygant BR, Lam CH, Burrow JN, Celio H, et al. Anodized nickel foam for oxygen evolution reaction in Fe-free and Unpurified alkaline electrolytes at high current densities. ACS Nano. 2021;**15**(2): 3468-3480. DOI: 10.1021/acsnano.0c10788

[16] Liang C, Zou P, Nairan A, Zhang Y, Liu J, Liu K, et al. Exceptional performance of hierarchical Ni–Fe oxyhydroxide@NiFe alloy nanowire Array electrocatalysts for large current density water splitting. Energy & Environmental Science. 2020;**13**(1):86-95. DOI: 10.1039/C9EE02388G

[17] Bediako DK, Surendranath Y, Nocera DG. Mechanistic studies of the oxygen evolution reaction mediated by a nickel–borate thin film electrocatalyst. Journal of the American Chemical Society. 2013;**135**(9):3662-3674. DOI: 10.1021/ja3126432

[18] Trześniewski BJ, Diaz-Morales O, Vermaas DA, Longo A, Bras W, Koper MTM, et al. In situ observation of active oxygen species in Fe-containing Ni-based oxygen evolution catalysts: The effect of pH on electrochemical activity. Journal of the American Chemical Society. 2015;**137**(48):15112-15121. DOI: 10.1021/jacs.5b06814

[19] Görlin M, Ferreira de Araújo J, Schmies H, Bernsmeier D, Dresp S, Gliech M, et al. Tracking catalyst redox states and reaction dynamics in Ni–Fe Oxyhydroxide oxygen evolution reaction electrocatalysts: The role of catalyst support and electrolyte pH. Journal of the American Chemical Society. 2017; **139**(5):2070-2082. DOI: 10.1021/jacs.6b12250

[20] Nguyen TN, Chen Z, Zeraati AS, Shiran HS, Sadaf SM, Kibria MG, et al. Catalyst regeneration via chemical oxidation enables long-term electrochemical carbon dioxide reduction. Journal of the American Chemical Society. 2022;**144**(29): 13254-13265. DOI: 10.1021/jacs.2c04081

[21] Stamatelos I, Scheepers F, Pasel J, Dinh C-T, Stolten D. Ternary Zn-Ce-Ag catalysts for selective and stable electrochemical CO_2 reduction at large-scale. Applied Catalysis B: Environment and Energy. 2024;**353**:124062. DOI: 10.1016/j.apcatb.2024.124062

[22] Gupta K, Bersani M, Darr JA. Highly efficient electro-reduction of CO_2 to formic acid by Nano-copper. Journal of Materials Chemistry A. 2016;**4**(36): 13786-13794. DOI: 10.1039/C6TA04874A

[23] Zhu S, Ren X, Li X, Niu X, Wang M, Xu S, et al. Core-Shell ZnO@Cu_2O as catalyst to enhance the electrochemical reduction of carbon dioxide to C_2 products. Catalysts. 2021;**11**(5):535. DOI: 10.3390/catal11050535

[24] Zhang T, Li Z, Zhang J, Wu J. Enhance CO_2-to-C_{2+} products yield through spatial management of CO transport in Cu/ZnO tandem electrodes. Journal of Catalysis. 2020;**387**:163-169. DOI: 10.1016/j.jcat.2020.05.002

[25] Qin B, Li Y, Fu H, Wang H, Chen S, Liu Z, et al. Electrochemical reduction of CO_2 into Tunable syngas production by regulating the crystal facets of earth-abundant Zn catalyst. ACS Applied Materials & Interfaces. 2018;**10**(24): 20530-20539. DOI: 10.1021/acsami.8b04809

[26] Peterson AA, Nørskov JK. Activity descriptors for CO_2 electroreduction to methane on transition-metal catalysts. Journal of Physical Chemistry Letters. 2012;**3**(2):251-258. DOI: 10.1021/jz201461p

[27] Liu X, Xiao J, Peng H, Hong X, Chan K, Nørskov JK. Understanding trends in electrochemical carbon dioxide reduction rates. Nature Communications. 2017;**8**(1):15438. DOI: 10.1038/ncomms15438

[28] Hansen HA, Varley JB, Peterson AA, Nørskov JK. Understanding trends in the electrocatalysts activity of metals and enzymes for CO_2 reduction to CO. Journal of Physical Chemistry Letters. 2013;**4**(3):388-392. DOI: 10.1021/jz3021155

[29] Cheng T, Xiao H, Goddard WAI. Reaction mechanisms for the electrochemical reduction of CO_2 to CO and Formate on the Cu(100) surface at 298 K from quantum mechanics free energy calculations with explicit water. Journal of the American Chemical Society. 2016;**138**(42):13802-13805. DOI: 10.1021/jacs.6b08534

[30] Liang S, Huang L, Gao Y, Wang Q, Liu B. Electrochemical reduction of CO_2 to CO over transition metal/N-doped carbon catalysts: The active sites and reaction mechanism. Advanced Science. 2021;**8**(24):2102886. DOI: 10.1002/advs.202102886

[31] Deng W, Zhang P, Seger B, Gong J. Unraveling the rate-limiting step of two-electron transfer electrochemical reduction of carbon dioxide. Nature Communications. 2022;**13**(1):803. DOI: 10.1038/s41467-022-28436-z

[32] Kari J, Olsen JP, Jensen K, Badino SF, Krogh KBRM, Borch K, et al. Sabatier principle for interfacial (heterogeneous) enzyme catalysis. ACS Catalysis. 2018;**8**(12):11966-11972. DOI: 10.1021/acscatal.8b03547

[33] Ďurovič M, Hnát J, Bouzek K. Electrocatalysts for the hydrogen evolution reaction in alkaline and neutral media. A comparative review. Journal of Power Sources. 2021;**493**:229708. DOI: 10.1016/j.jpowsour.2021.229708

[34] Jakšić JM, Vojnović MV, Krstajić NV. Kinetic analysis of hydrogen evolution at Ni–Mo alloy electrodes. Electrochimica Acta. 2000;**45**(25):4151-4158. DOI: 10.1016/S0013-4686(00)00549-1

[35] Conway BE, Tilak BV. Interfacial processes involving electrocatalytic evolution and oxidation of H_2, and the role of chemisorbed H. Electrochimica Acta. 2002;**47**(22):3571-3594. DOI: 10.1016/S0013-4686(02)00329-8

[36] Handoko AD, Wei F, Jenndy, Yeo BS, Seh ZW. Understanding heterogeneous electrocatalysts carbon dioxide reduction through operando techniques. Nature Catalysis. 2018;**1**(12):922-934. DOI: 10.1038/s41929-018-0182-6

[37] Marrero TR, Mason EA. Gaseous diffusion coefficients. Journal of Physical and Chemical Reference Data. 2009;**1**(1):3-118. DOI: 10.1063/1.3253094

[38] Casebolt DiDomenico R, Levine K, Reimanis L, Abruña HD, Hanrath T. Mechanistic insights into the formation of CO and C_2 products in electrochemical CO_2 reduction—the role of sequential charge transfer and chemical reactions. ACS Catalysis. 2023;**13**(7):4938-4948. DOI: 10.1021/acscatal.2c06043

[39] Zheng W, Wang D, Zhang Y, Zheng S, Yang B, Li Z, et al. Promoting

industrial-level CO_2 electroreduction kinetics via accelerating proton feeding on a metal-free aerogel electrocatalyst. Nano Energy. 2023;**105**:107980. DOI: 10.1016/j.nanoen.2022.107980

[40] Yang W, Xue Z, Yang J, Xian J, Liu Q, Fan Y, et al. Fe nanoparticles embedded in N-doped porous carbon for enhanced electrocatalysts CO_2 reduction and Zn-CO_2 battery. Chinese Journal of Catalysis. 2023;**48**:185-194. DOI: 10.1016/S1872-2067(23)64415-8

[41] Rossmeisl J, Qu Z-W, Zhu H, Kroes G-J, Nørskov JK. Electrolysis of water on oxide surfaces. Journal of Electroanalytical Chemistry. 2007;**607**(1):83-89. DOI: 10.1016/j.jelechem.2006.11.008

[42] Medford AJ, Vojvodic A, Hummelshøj JS, Voss J, Abild-Pedersen F, Studt F, et al. From the Sabatier principle to a predictive theory of transition-metal heterogeneous catalysis. Journal of Catalysis. 2015;**328**:36-42. DOI: 10.1016/j.jcat.2014.12.033

[43] Rong X, Parolin J, Kolpak AM. A fundamental relationship between reaction mechanism and stability in metal oxide catalysts for oxygen evolution. ACS Catalysis. 2016;**6**(2):1153-1158. DOI: 10.1021/acscatal.5b02432

[44] Fei H, Dong J, Feng Y, Allen CS, Wan C, Volosskiy B, et al. General synthesis and definitive structural identification of MN4C4 single-atom catalysts with Tunable electrocatalysts activities. Nature Catalysis. 2018;**1**(1):63-72. DOI: 10.1038/s41929-017-0008-y

[45] Hwang J, Rao RR, Giordano L, Katayama Y, Yu Y, Shao-Horn Y. Perovskites in catalysis and Electrocatalysis. Science. 2017;**358**(6364):751-756. DOI: 10.1126/science.aam7092

[46] Mefford JT, Rong X, Abakumov AM, Hardin WG, Dai S, Kolpak AM, et al. Water electrolysis on La1−xSrxCo$O_{3-\delta}$ perovskite electrocatalysts. Nature Communications. 2016;7(1):11053. DOI: 10.1038/ncomms11053

[47] Nolan M, Iwaszuk A, Lucid AK, Carey JJ, Fronzi M. Design of Novel visible light active photocatalyst materials: Surface modified TiO_2. Advanced Materials. 2016;**28**(27):5425-5446. DOI: 10.1002/adma.201504894

[48] Liu K, Ma M, Wu L, Valenti M, Cardenas-Morcoso D, Hofmann JP, et al. Electronic effects determine the selectivity of planar Au–Cu bimetallic thin films for electrochemical CO_2 reduction. ACS Applied Materials & Interfaces. 2019;**11**(18):16546-16555. DOI: 10.1021/acsami.9b01553

[49] Jeyachandran N, Yuan W, Giordano C. Cutting-edge electrocatalysts for CO2RR. Molecules. 2023;**28**(8):3504. DOI: 10.3390/molecules28083504

[50] Ringe S, Morales-Guio CG, Chen LD, Fields M, Jaramillo TF, Hahn C, et al. Double layer charging driven carbon dioxide adsorption limits the rate of electrochemical carbon dioxide reduction on gold. Nature Communications. 2020;**11**(1):33. DOI: 10.1038/s41467-019-13777-z

[51] Mosali VSS, Bond AM, Zhang J. Alloying strategies for tuning product selectivity during electrochemical CO_2 reduction over Cu. Nanoscale. 2022;**14**(42):15560-15585. DOI: 10.1039/D2NR03539A

[52] Xu Y, Li C, Xiao Y, Wu C, Li Y, Li Y, et al. Tuning the selectivity of liquid products of CO2RR by Cu–Ag alloying.

ACS Applied Materials & Interfaces. 2022;**14**(9):11567-11574. DOI: 10.1021/acsami.2c00593

[53] Wu Q, Gao Q, Wang X, Qi Y, Shen L, Tai X, et al. Boosting electrocatalysts performance via electronic structure regulation for acidic oxygen evolution. iScience. 2024;**27**(1):108738. DOI: 10.1016/j.isci.2023.108738

[54] Lee S, Shin Y, Yeom K, Shim J, Sung Y-E. An overview of design strategies and recent advancements in complex 3d transition metal-based electrocatalysts for alkaline oxygen evolution reaction. Advances in Industrial and Engineering Chemistry. 2025;**1**(1):2. DOI: 10.1007/s44405-025-00001-4

[55] Kim U, Mun J, Koo D, Seo J, Choi Y, Lee G, et al. Catalytic centers with multiple oxidation states: A strategy for breaking the overpotential ceiling from the linear scaling relation in oxygen evolution. Journal of Materials Chemistry A. 2022;**10**(43):23079-23086. DOI: 10.1039/D2TA05409D

[56] Sun Y, Chen G, Xi S, Xu ZJ. Catalytically influential features in transition metal oxides. ACS Catalysis. 2021;**11**(22):13947-13954. DOI: 10.1021/acscatal.1c04393

[57] Marquez RA, Kalokowski E, Espinosa M, Bender JT, Son YJ, Kawashima K, et al. Transition metal incorporation: Electrochemical, structure, and chemical composition effects on nickel oxyhydroxide oxygen-evolution electrocatalysts. Energy & Environmental Science. 2024;**17**(5):2028-2045. DOI: 10.1039/D3EE03617K

[58] Song F, Bai L, Moysiadou A, Lee S, Hu C, Liardet L, et al. Transition metal oxides as electrocatalysts for the oxygen evolution reaction in alkaline solutions: An application-inspired renaissance. Journal of the American Chemical Society. 2018;**140**(25):7748-7759. DOI: 10.1021/jacs.8b04546

[59] Stamatelos I, da Silva GTST, Ribeiro C, Shviro M. Exploring Heterostructures of D-block metal oxides coupled to ZnO for the electrochemical reduction of CO_2. ACS Applied Energy Materials. 2023;**6**(22):11510-11520. DOI: 10.1021/acsaem.3c01791

[60] Whang HS, Lim J, Choi MS, Lee J, Lee H. Heterogeneous catalysts for catalytic CO_2 conversion into value-added chemicals. BMC Chemical Engineering. 2019;**1**(1):9. DOI: 10.1186/s42480-019-0007-7

[61] Ma M, Trześniewski BJ, Xie J, Smith WA. Selective and efficient reduction of carbon dioxide to carbon monoxide on oxide-derived nanostructured silver electrocatalysts. Angewandte Chemie International Edition. 2016;**55**(33):9748-9752. DOI: 10.1002/anie.201604654

[62] Guo Z, Wang T, Liu H, Jia X, Zhang D, Wei L, et al. Electrochemical CO_2 reduction on SnO: Insights into C1 product dynamic distribution and reaction mechanisms. ACS Catalysis. 2025;**15**(4):3173-3183. DOI: 10.1021/acscatal.4c07987

[63] Stamatelos I, Dinh C-T, Lehnert W, Shviro M. Zn-based catalysts for selective and stable electrochemical CO_2 reduction at high current densities. ACS Applied Energy Materials. 2022;**5**(11):13928-13938. DOI: 10.1021/acsaem.2c02557

[64] Dattila F, García-Muelas R, López N. Active and selective ensembles in oxide-derived copper catalysts for CO_2 reduction. ACS Energy Letters. 2020;**5**(10):3176-3184. DOI: 10.1021/acsenergylett.0c01777

[65] Onoh EU, Stamatelos I, Dubale AA, Pasel J, Ayalneh Tiruye G. Selective electro-reduction of CO_2 into methane and formic acid using efficient bimetallic and bimetallic oxide electrocatalysts in liquid-fed Electrolyzers. Journal of Power Sources. 2025;**633**:236393. DOI: 10.1016/j.jpowsour.2025.236393

[66] He H, Xia D, Yu X, Wu J, Wang Y, Wang L, et al. Pd-SnO_2 Interface enables synthesis of syngas with controllable H_2/CO ratios by electrocatalysts reduction of CO_2. Applied Catalysis B: Environmental. 2022;**312**:121392. DOI: 10.1016/j.apcatb.2022.121392

[67] Fu HQ, Liu J, Bedford NM, Wang Y, Sun JW, Zou Y, et al. Synergistic Cr2O3@Ag Heterostructure enhanced electrocatalysts CO_2 reduction to CO. Advanced Materials. 2022;**34**(29): 2202854. DOI: 10.1002/adma.202202854

[68] Yang Y, Guo M, Zhao F. Cr_2O_3 promoted In_2O_3 catalysts for CO_2 hydrogenation to methanol. ChemPhysChem. 2024;**25**:e202300530. DOI: 10.1002/cphc.202300530

[69] Bao F, Kemppainen E, Dorbandt I, Xi F, Bors R, Maticiuc N, et al. Host, suppressor, and promoter—The roles of Ni and Fe on oxygen evolution reaction activity and stability of NiFe alloy thin films in alkaline media. ACS Catalysis. 2021;**11**(16):10537-10552. DOI: 10.1021/acscatal.1c01190

[70] Yu M, Moon G, Bill E, Tüysüz H. Optimizing Ni–Fe oxide electrocatalysts for oxygen evolution reaction by using hard templating as a toolbox. ACS Applied Energy Materials. 2019;**2**(2): 1199-1209. DOI: 10.1021/acsaem.8b01769

[71] Avcı ÖN, Sementa L, Fortunelli A. Mechanisms of the oxygen evolution reaction on $NiFe_2$ O4 and $CoFe_2$ O_4 inverse-spinel oxides. ACS Catalysis. 2022;**12**(15):9058-9073. DOI: 10.1021/acscatal.2c01534

[72] Jang MJ, Yang J, Lee J, Park YS, Jeong J, Park SM, et al. Superior performance and stability of anion exchange membrane water electrolysis: pH-controlled copper cobalt oxide nanoparticles for the oxygen evolution reaction. Journal of Materials Chemistry A. 2020;**8**(8):4290-4299. DOI: 10.1039/C9TA13137J

[73] Mu G, Zeng Y, Zheng Y, Cao Y, Liu F, Liang S, et al. Oxygen vacancy defects engineering on Cu-doped Co_3O_4 for promoting effective COS hydrolysis. Green Energy & Environment. 2023;**8**(3):831-841. DOI: 10.1016/j.gee.2021.11.001

[74] Zhang L, Li Y, Peng J, Peng K. Bifunctional $NiCo_2O_4$ porous nanotubes electrocatalyst for overall water-splitting. Electrochimica Acta. 2019;**318**: 762-769. DOI: 10.1016/j.electacta.2019.06.128

[75] Jiang B, Wan Z, Kang Y, Guo Y, Henzie J, Na J, et al. Auto-programmed synthesis of metallic aerogels: Core-Shell Cu@Fe@Ni aerogels for efficient oxygen evolution reaction. Nano Energy. 2021; **81**:105644. DOI: 10.1016/j.nanoen.2020.105644

[76] Xiao J, Liu S, Sui P-F, Xu C, Gong L, Zeng H, et al. In-situ generated hydroxides realize near-Unity CO selectivity for electrochemical CO_2 reduction. Chemical Engineering Journal. 2022;**433**:133785. DOI: 10.1016/j.cej.2021.133785

[77] Dinh C-T, Burdyny T, Kibria MG, Seifitokaldani A, Gabardo CM, García de Arquer FP, et al. CO_2 electroreduction to ethylene via hydroxide-mediated copper catalysis at an abrupt Interface. Science.

2018;**360**(6390):783-787. DOI: 10.1126/science.aas9100

[78] Lee JH, Jang W, Lee H, Oh D, Noh WY, Kim KY, et al. Tuning CuMgAl-layered double hydroxide nanostructures to achieve CH_4 and C_{2+} product selectivity in CO_2 electroreduction. Nano Letters. 2024;**24**(30):9322-9330. DOI: 10.1021/acs.nanolett.4c02233

[79] Senanayake SD, Stacchiola D, Evans J, Estrella M, Barrio L, Pérez M, et al. Probing the reaction intermediates for the water–gas shift over inverse CeOx/Au (1 1 1) catalysts. Journal of Catalysis. 2010;**271**(2):392-400. DOI: 10.1016/j.jcat.2010.02.024

[80] Henderson MA, Perkins CL, Engelhard MH, Thevuthasan S, Peden CHF. Redox properties of water on the oxidized and reduced surfaces of CeO_2(111). Surface Science. 2003;**526**(1):1-18. DOI: 10.1016/S0039-6028(02)02657-2

[81] Sun M, Staykov A, Yamauchi M. Understanding the roles of hydroxide in CO_2 electroreduction on a Cu electrode for achieving variable selectivity. ACS Catalysis. 2022;**12**(24):14856-14863. DOI: 10.1021/acscatal.2c03650

[82] Gao L, Cui X, Sewell CD, Li J, Lin Z. Recent advances in activating surface reconstruction for the high-efficiency oxygen evolution reaction. Chemical Society Reviews. 2021;**50**(15):8428-8469. DOI: 10.1039/D0CS00962H

[83] Lee WH, Han MH, Ko Y-J, Min BK, Chae KH, Oh H-S. Electrode reconstruction strategy for oxygen evolution reaction: Maintaining Fe-CoOOH phase with intermediate-spin state during electrolysis. Nature Communications. 2022;**13**(1):605. DOI: 10.1038/s41467-022-28260-5

[84] He Z-D, Tesch R, Eslamibidgoli MJ, Eikerling MH, Kowalski PM. Low-spin state of Fe in Fe-doped NiOOH electrocatalysts. Nature Communications. 2023;**14**(1):3498. DOI: 10.1038/s41467-023-38978-5

[85] Chen JYC, Dang L, Liang H, Bi W, Gerken JB, Jin S, et al. Operando analysis of NiFe and Fe oxyhydroxide electrocatalysts for water oxidation: Detection of Fe_{4+} by Mössbauer spectroscopy. Journal of the American Chemical Society. 2015;**137**(48):15090-15093. DOI: 10.1021/jacs.5b10699

[86] Tsyganok A, Ghigna P, Minguzzi A, Naldoni A, Murzin V, Caliebe W, et al. Operando X-ray absorption spectroscopy (XAS) observation of Photoinduced oxidation in FeNi (oxy) hydroxide Overlayers on hematite (α-Fe_2O_3) Photoanodes for solar water splitting. Langmuir. 2020;**36**(39):11564-11572. DOI: 10.1021/acs.langmuir.0c02065

[87] Aigul S, Enkhbayar E, Gaur A, Han H. Cr-doped tri-metallic Nano prism catalyst for efficient alkaline and seawater splitting. Journal of Crystal Growth. 2025;**649**:127928. DOI: 10.1016/j.jcrysgro.2024.127928

[88] Babu SP, Falch A. Recent developments on Cr-based electrocatalysts for the oxygen evolution reaction in alkaline media. ChemCatChem. 2022;**14**(15):e202200364. DOI: 10.1002/cctc.202200364

[89] Cao F, Li M, Hu Y, Wu X, Li X, Meng X, et al. Kinetically accelerated oxygen evolution reaction in metallic (oxy)hydroxides enabled by Cr-dopant and Heterostructure. Chemical Engineering Journal. 2023;**472**:144970. DOI: 10.1016/j.cej.2023.144970

[90] Zhou D, Li P, Lin X, McKinley A, Kuang Y, Liu W, et al. Layered double hydroxide-based electrocatalysts for the oxygen evolution reaction: Identification and tailoring of active sites, and Superaerophobic Nanoarray electrode assembly. Chemical Society Reviews. 2021;**50**(15):8790-8817. DOI: 10.1039/D1CS00186H

[91] Lei H, Ma L, Wan Q, Tan S, Yang B, Wang Z, et al. Promoting surface reconstruction of NiFe layered double hydroxide for enhanced oxygen evolution. Advanced Energy Materials. 2022;**12**(48):2202522. DOI: 10.1002/aenm.202202522

[92] Wu L, Yu L, Xiao X, Zhang F, Song S, Chen S, et al. Recent advances in self-supported layered double hydroxides for oxygen evolution reaction. Research. 2020;**2020**. DOI: 10.34133/2020/3976278

[93] Song F, Hu X. Exfoliation of layered double hydroxides for enhanced oxygen evolution catalysis. Nature Communications. 2014;**5**(1):4477. DOI: 10.1038/ncomms5477

[94] Wan X, Song Y, Zhou H, Shao M. Layered double hydroxides for oxygen evolution reaction towards efficient hydrogen generation. Energy Material Advances. 2022;**2022**. DOI: 10.34133/2022/9842610

[95] Qu J, Dong Y, Zhang T, Zhao C, Wei L, Guan X. Impact of bimetallic synergies on Mo-doping NiFeOOH: Insights into enhanced OER activity and reconstructed electronic structure. Frontiers in Energy. 2024;**18**(6):850-862. DOI: 10.1007/s11708-024-0960-6

[96] He Z, Zhang J, Gong Z, Lei H, Zhou D, Zhang N, et al. Activating lattice oxygen in NiFe-based (oxy)hydroxide for water electrolysis. Nature Communications. 2022;**13**(1):2191. DOI: 10.1038/s41467-022-29875-4

[97] Krivina RA, Ou Y, Xu Q, Twight LP, Stovall TN, Boettcher SW. Oxygen Electrocatalysis on mixed-metal oxides/Oxyhydroxides: From fundamentals to membrane Electrolyzer technology. Accounts of Materials Research. 2021;**2**(7):548-558. DOI: 10.1021/accountsmr.1c00087

[98] Iijima G, Inomata T, Yamaguchi H, Ito M, Masuda H. Role of a hydroxide layer on Cu electrodes in electrochemical CO_2 reduction. ACS Catalysis. 2019;**9**(7):6305-6319. DOI: 10.1021/acscatal.9b00896

[99] Chen Y, Li CW, Kanan MW. Aqueous CO_2 reduction at very low Overpotential on oxide-derived Au nanoparticles. Journal of the American Chemical Society. 2012;**134**(49):19969-19972. DOI: 10.1021/ja309317u

[100] Löffler M, Mayrhofer KJJ, Katsounaros I. Oxide reduction precedes carbon dioxide reduction on oxide-derived copper electrodes. Journal of Physical Chemistry C. 2021;**125**(3):1833-1838. DOI: 10.1021/acs.jpcc.0c09107

[101] Mattarozzi F, van der Willige N, Gulino V, Keijzer C, van de Poll RCJ, Hensen EJM, et al. Oxide-derived silver nanowires for CO_2 electrocatalysts reduction to CO. CheCatChem. 2023;**15**:e202300792. DOI: 10.1002/cctc.202300792

[102] Hu H, Tang Y, Hu Q, Wan P, Dai L, Yang XJ. In-situ grown Nanoporous Zn-Cu catalysts on Brass foils for enhanced electrochemical reduction of carbon dioxide. Applied Surface Science. 2018;**445**:281-286. DOI: 10.1016/j.apsusc.2018.03.146

[103] Eilert A, Cavalca F, Roberts FS, Osterwalder J, Liu C, Favaro M, et al.

Subsurface oxygen in oxide-derived copper electrocatalysts for carbon dioxide reduction. Journal of Physical Chemistry Letters. 2017;**8**(1): 285-290. DOI: 10.1021/acs.jpclett.6b02273

[104] Asiri AM, Gao J, Khan SB, Alamry KA, Marwani HM, Khan MSJ, et al. Revisiting the impact of morphology and oxidation state of Cu on CO_2 reduction using electrochemical flow cell. Journal of Physical Chemistry Letters. 2022;**13** (1):345-351. DOI: 10.1021/acs.jpclett.1c03957

[105] Behnaz Varandili S, Stoian D, Vavra J, Rossi K, Pankhurst R, Guntern Y, et al. Elucidating the structure-dependent selectivity towards methane and ethanol of CuZn in the CO_2 electroreduction using tailored Cu/ZnO Precatalysts. Chemical Science. 2021;**12**:14484-14493. DOI: 10.1039/D1SC04271H

[106] Torelli DA, Francis SA, Crompton JC, Javier A, Thompson JR, Brunschwig BS, et al. Nickel–gallium-catalyzed electrochemical reduction of CO_2 to highly reduced products at low overpotentials. ACS Catalysis. 2016;**6**(3): 2100-2104. DOI: 10.1021/acscatal.5b02888

[107] Zhan C, Dattila F, Rettenmaier C, Herzog A, Herran M, Wagner T, et al. Key intermediates and Cu active sites for CO_2 electroreduction to ethylene and ethanol. Nature Energy. 2024;**9**(12): 1485-1496. DOI: 10.1038/s41560-024-01633-4

[108] Wei D, Wang Y, Dong C-L, Nga TT, Shi Y, Wang J, et al. Surface adsorbed hydroxyl: A double-edged sword in electrochemical CO_2 reduction over oxide-derived copper. Angewandte Chemie International Edition. 2023;**62** (31):e202306876. DOI: 10.1002/anie.202306876

[109] Luo M, Wang Z, Li YC, Li J, Li F, Lum Y, et al. Hydroxide promotes carbon dioxide electroreduction to ethanol on copper via tuning of adsorbed hydrogen. Nature Communications. 2019;**10**(1):5814. DOI: 10.1038/s41467-019-13833-8

[110] Dionigi F, Zeng Z, Sinev I, Merzdorf T, Deshpande S, Lopez MB, et al. In-situ structure and catalytic mechanism of NiFe and CoFe layered double hydroxides during oxygen evolution. Nature Communications. 2020;**11**(1):2522. DOI: 10.1038/s41467-020-16237-1

[111] Wang Y, Liu S, Qin Y, Zhao Y, Liu L, Zhang D, et al. Chromium promotes phase transformation to active Oxyhydroxide for efficient oxygen evolution. ACS Catalysis. 2024;**14**(18): 13759-13767. DOI: 10.1021/acscatal.4c03974

[112] Wang Y, Chen C, Xiong X, Skaanvik SA, Zhang Y, Bøjesen ED, et al. In situ tracking of water oxidation generated nanoscale dynamics in layered double hydroxides nanosheets. Journal of the American Chemical Society. 2024; **146**(25):17032-17040. DOI: 10.1021/jacs.4c01035

[113] Wang Z, Goddard WA, Xiao H. Potential-dependent transition of reaction mechanisms for oxygen evolution on layered double hydroxides. Nature Communications. 2023;**14**(1): 4228. DOI: 10.1038/s41467-023-40011-8

[114] Gao MY, Sun CB, Lei H, Zeng JR, Zhang QB. Nitrate-induced and in situ electrochemical activation synthesis of oxygen deficiencies-rich nickel/nickel (oxy)hydroxide hybrid films for enhanced electrocatalysts water splitting. Nanoscale. 2018;**10**(37):17546-17551. DOI: 10.1039/C8NR06459H

[115] Jiang Q, Wang S, Zhang C, Sheng Z, Zhang H, Feng R, et al. Active oxygen species mediate the iron-promoting Electrocatalysis of oxygen evolution reaction on metal Oxyhydroxides. Nature Communications. 2023;**14**(1): 6826. DOI: 10.1038/s41467-023-42646-z

[116] Ping X, Liu Y, Zheng L, Song Y, Guo L, Chen S, et al. Locking the lattice oxygen in RuO_2 to stabilize highly active Ru sites in acidic water oxidation. Nature Communications. 2024;**15**(1):2501. DOI: 10.1038/s41467-024-46815-6

[117] Ferreira de Araújo J, Dionigi F, Merzdorf T, Oh H-S, Strasser P. Evidence of Mars-van-Krevelen mechanism in the electrochemical oxygen evolution on Ni-based catalysts. Angewandte Chemie International Edition. 2021;**60**(27):14981-14988. DOI: 10.1002/anie.202101698

[118] Arán-Ais RM, Scholten F, Kunze S, Rizo R, Roldan Cuenya B. The role of in situ generated morphological motifs and Cu(i) species in C_{2+} product selectivity during CO_2 pulsed electroreduction. Nature Energy. 2020;**5**(4):317-325. DOI: 10.1038/s41560-020-0594-9

[119] Reichert MA, Piqué O, Parada AW, Katsounaros I, Calle-Vallejo F. Mechanistic insight into electrocatalytic glyoxal reduction on copper and its relation to CO_2 reduction. Chemical Science. 2022;**13**(37):11205-11214. DOI: 10.1039/D2SC03527H

[120] Cheng D, Zhao Z-J, Zhang G, Yang P, Li L, Gao H, et al. The nature of active sites for carbon dioxide electroreduction over oxide-derived copper catalysts. Nature Communications. 2021;**12**(1): 395. DOI: 10.1038/s41467-020-20615-0

[121] Kim C, Bui JC, Luo X, Cooper JK, Kusoglu A, Weber AZ, et al. Tailored catalyst microenvironments for CO_2 electroreduction to multicarbon products on copper using bilayer ionomer coatings. Nature Energy. 2021;**6** (11):1026-1034. DOI: 10.1038/s41560-021-00920-8

[122] Zeng J, Castellino M, Fontana M, Sacco A, Monti NBD, Chiodoni A, et al. Electrochemical reduction of CO_2 with good efficiency on a nanostructured Cu-Al catalyst. Frontiers in Chemistry. 2022: 10. DOI: 10.3389/fchem.2022.931767

[123] Li F, Thevenon A, Rosas-Hernández A, Wang Z, Li Y, Gabardo CM, et al. Molecular tuning of CO_2-to-ethylene conversion. Nature. 2020;**577**(7791):509-513. DOI: 10.1038/s41586-019-1782-2

[124] Zong X, Zhang J, Zhang J, Luo W, Züttel A, Xiong Y. Synergistic Cu/CeO_2 carbon nanofiber catalysts for efficient CO_2 electroreduction. Electrochemistry Communications. 2020;**114**:106716. DOI: 10.1016/j.elecom.2020.106716

[125] Dongare S, Singh N, Bhunia H. Oxide-derived Cu-Zn nanoparticles supported on N-doped graphene for electrochemical reduction of CO_2 to ethanol. Applied Surface Science. 2021; **556**:149790. DOI: 10.1016/j.apsusc.2021.149790

[126] You S, Xiao J, Liang S, Xie W, Zhang T, Li M, et al. Doping engineering of Cu-based catalysts for electrocatalysts CO_2 reduction to multi-carbon products. Energy and Environmental Science. 2024;**17**(16):5795-5818. DOI: 10.1039/D4EE01325E

[127] Knöppel J, Möckl M, Escalera-López D, Stojanovski K, Bierling M, Böhm T, et al. On the limitations in assessing stability of oxygen evolution catalysts using aqueous model electrochemical cells. Nature Communications. 2021;**12** (1):2231. DOI: 10.1038/s41467-021-22296-9

[128] Zhang D, Wu Q, Wu L, Cheng L, Huang K, Chen J, et al. Optimal electrocatalyst design strategies for acidic oxygen evolution. Advanced Science. 2024;**11**(38):2401975. DOI: 10.1002/advs.202401975

[129] Zhang C, Wang F, Batool M, Xiong B, Yang H. Phase transition of $SrCo_{0.9}Fe_{0.1}O_3$ electrocatalysts and their effects on oxygen evolution reaction. SusMat. 2022;**2**(4):445-455. DOI: 10.1002/sus2.72

[130] Yin X, Hua Y, Gao Z. Two-dimensional materials for high-performance oxygen evolution reaction: Fundamentals, recent progress, and improving strategies. Renewables. 2023;**1**(2):190-226. DOI: 10.31635/renewables.023.202200003

[131] Zhang Y, Fu Q, Song B, Xu P. Regulation strategy of transition metal oxide-based electrocatalysts for enhanced oxygen evolution reaction. Accounts of Materials Research. 2022;**3**(10):1088-1100. DOI: 10.1021/accountsmr.2c00161

[132] Park W, Chung DY. Activity–stability relationships in oxygen evolution reaction. ACS Materials Au. 2025;**5**(1):1-10. DOI: 10.1021/acsmaterialsau.4c00086

Section 2

New Methods of Electrode Preparation for Electrocatalytic Applications

Chapter 3

Perspective Chapter: Rotating Disc Slurry Electrodeposition – An Innovative Approach to Synthesising PGM, Non-PGM, and Alloy Nanoparticles for Oxygen Reduction Reaction Catalysis

Suzatra Chatterjee, Abayomi Omoogun, Ruma Rani Paul, Subhajit Bhunia and Carlos R. Cabrera

Abstract

The rotating disc slurry electrodeposition (RoDSE) technique is a cutting-edge method for synthesising advanced catalysts tailored for the oxygen reduction reaction (ORR). ORR, a critical process in fuel cells and energy devices, involves two key pathways: the four-electron pathway ($O_2 \rightarrow H_2O$) for efficient energy conversion and the two-electron pathway ($O_2 \rightarrow H_2O_2$), often detrimental in fuel cells. RoDSE enables the precise synthesis of nanoparticles on different supports, including single metals (e.g., Pt, Ag, Pd) and alloys (e.g., Pt-Ni, Ag-Pd, Pt-Fe-Co). This scalable and eco-friendly technique enhances catalytic activity and durability, advancing cost-effective, high-performance ORR catalysts for clean energy applications.

Keywords: RoDSE, electrodeposition, catalysis, ORR, alkaline medium, scalable

1. Introduction

The oxygen reduction reaction (ORR) [1] is a crucial electrochemical process in energy conversion and storage systems, including fuel cells [2], metal-air batteries [3], and hydrogen peroxide production. ORR is vital for the efficiency and lifespan of these devices because it often exhibits slow kinetics, necessitating the use of effective electrocatalysts to accelerate the reaction. The preferred four-electron pathway ($O_2 + 4H^+ + 4e^- \rightarrow 2H_2O$ in acidic media or $O_2 + 2H_2O + 4e^- \rightarrow 4OH^-$ in alkaline media) [4] allows for optimal energy conversion, while the two-electron pathway ($O_2 + 2H^+ + 2e^- \rightarrow H_2O_2$) results in peroxide formation [5], which diminishes efficiency and harms fuel cell membranes. The oxygen reduction reaction (ORR) exhibits significant differences in alkaline and acidic environments [6], primarily in terms of reaction mechanisms, kinetics, and catalyst performance. In acidic media, the reaction involves the direct combination of protons (H^+) with oxygen. At the same time, in alkaline conditions, water molecules act as proton donors, resulting in the formation of hydroxide ions (OH^-). The alkaline environment generally facilitates more favourable ORR kinetics due to improved activation of the peroxide intermediate. Catalyst behaviour can vary depending on the pH, with some materials showing superior performance in alkaline conditions.

To address these issues, electrocatalysts have been developed to enhance ORR kinetics and reduce overpotentials. Historically, platinum-group metals (PGMs), especially platinum [7] and its alloys [8], have served as the most efficient ORR catalysts. Nonetheless, their high cost and limited supply make it essential to investigate alternative materials, such as non-PGM catalysts [9], transition-metal alloys [10], carbon-based substances [11], metal-incorporated polymers [12], and covalent organic frameworks [13]. Traditional synthesis techniques—including chemical reduction [14], hydrothermal treatments [15], and solvothermal processes—often result in poor nanoparticle dispersion, uncontrolled size distribution, and agglomeration, which diminishes catalytic efficacy.

The rotating disc slurry electrodeposition (RoDSE) [16] technique presents a breakthrough in electrocatalyst synthesis, addressing these limitations by enabling uniform nanoparticle deposition on conductive carbon supports. This method enables precise control over metal composition, particle size, and distribution, thereby enhancing both catalytic activity and stability. RoDSE has been successfully applied to synthesise monometallic (Pt [17], Ag [18], Pd [19], Fe [5]) and bimetallic/alloy (Pt-Ni [20], Pt-Fe [20], Pt-Co [20], Ag-Pd [19]) catalysts, offering cost-effective and scalable solutions for high-performance ORR applications.

Unlike conventional electrodeposition methods, where metal ions are directly reduced on a stationary electrode, RoDSE utilises a rotating disc electrode (RDE) immersed in a slurry containing carbon support and metal precursors. This setup ensures uniform mass transport and electrochemical reduction, leading to highly dispersed nanoparticles with precise size and composition control. The process begins by preparing a slurry, which involves suspending a functionalised support material, such as carbon black, in an electrolyte solution. This slurry is placed in a three-electrode cell, with a rotating disc electrode (RDE) as the working electrode. The key to the RoDSE method lies in applying a constant potential to the RDE as it rotates within the slurry, creating a controlled hydrodynamic environment that ensures uniform accessibility of the electrolyte to the electrode surface. During this process, metal precursors in the electrolyte are reduced and deposited onto the suspended support material. After deposition, the slurry is filtered and rinsed to collect the catalyst material.

RoDSE offers several advantages over traditional synthesis techniques. It allows bulk catalyst synthesis, making it scalable for industrial applications. Since the carbon support is suspended in the electrolyte rather than directly attached to an electrode, deposition occurs uniformly, preventing aggregation and allowing for a high surface area of active sites. The method also provides better control over particle size and distribution, which is crucial for enhancing catalytic performance in the oxygen reduction reaction (ORR). Additionally, RoDSE minimises material waste and can be optimised for low-cost, high-efficiency electrocatalysts, making it a promising alternative for next-generation fuel cells and sustainable energy technologies.

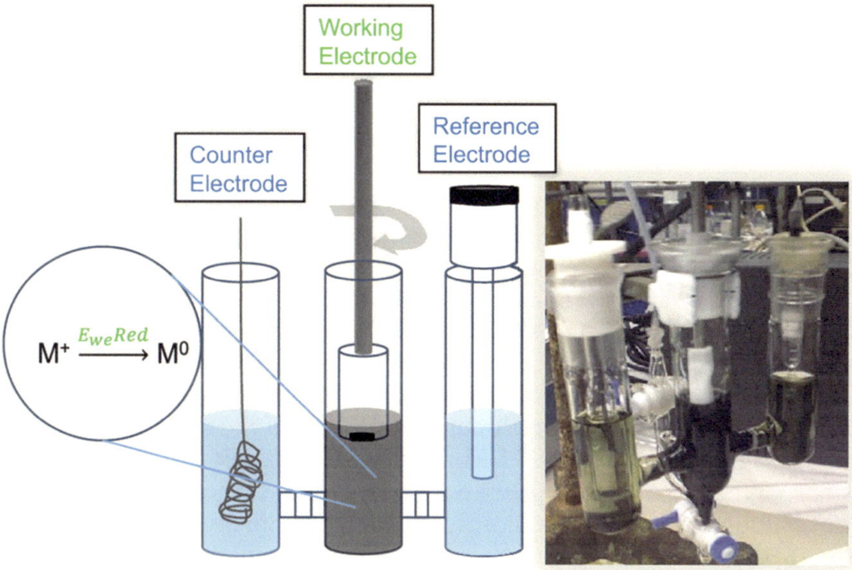

Figure 1.
RoDSE methodology scheme [20] (left) using a three-electrode electrochemical setup. Right, a picture of the synthesis setup. (Reproduced with permission of the publisher).

The schematic and accompanying photograph (**Figure 1**) depict the three-electrode rotating disc slurry electrodeposition (RoDSE) setup used to deposit metal nanoparticles onto a suspended carbon support. At the centre is the rotating disc working electrode, immersed in a stirred slurry of conductive carbon particles and metal-ion precursors; as the electrode spins, fresh precursor solution is drawn towards its surface, where the applied potential (E(we)) reduces M^{n+} ions to metallic M^0, nucleating uniformly on the carbon. To the left sits the counter electrode, which completes the circuit by balancing the working-electrode reaction, while the reference electrode on the right provides a stable potential reference. The inset chemical equation emphasises the electroreduction step ($M^{n+} \rightarrow M^0$) at the working electrode. In the photograph, one can see the custom-built cell with gas lines for argon purging, the rotating shaft entering through a glass frit, and the separated compartments for the reference and counter electrodes, underscoring how RoDSE decouples support contact from electrodeposition, ensuring highly dispersed nanoparticle growth across the entire carbon slurry.

Compared to traditional electrodeposition, where metal deposition occurs directly on an electrode surface, RoDSE allows deposition on unsupported carbon materials suspended in the slurry, ensuring a more homogeneous nanoparticle distribution. In conventional electrodeposition, agglomeration and uneven deposition are common issues, especially when scaling up the process. RoDSE overcomes these limitations by preventing direct electrode-substrate interactions, allowing for bulk synthesis of catalysts, making it more scalable for industrial applications.

On the other hand, chemical reduction methods, such as the polyol process or sodium borohydride ($NaBH_4$) reduction, involve reducing metal precursors in solution with chemical agents. While these methods are widely used for nanoparticle synthesis, they suffer from poor size control, batch-to-batch variations, and excessive use of stabilisers, which can block active catalytic sites. Moreover, chemical reduction techniques often result in random metal deposition and necessitate post-synthesis treatments to enhance catalytic activity and stability.

2. Role of RoDSE method in ORR catalyst development

Rotating disc slurry electrodeposition (RoDSE) is a novel electrochemical technique developed to synthesise finely dispersed catalysts on high-surface-area supports. In a traditional electrodeposition, one might deposit metal onto a flat rotating disc electrode (RDE); in RoDSE, the electrodeposition is carried out in a slurry containing the support material (e.g., carbon powder) suspended in the electrolyte. The rotating electrode stirs the slurry and enables metal ions to electrodeposit onto the numerous support particles in suspension. This method effectively "plates" catalyst nanoparticles onto the support throughout the bulk of the slurry, rather than just on a single electrode's surface. The RDE's rotation creates a well-defined, controlled mass transport of reactants to the electrode surface. This is crucial for quantitative electrochemical studies. The rotation ensures that the entire disc surface is uniformly accessible to the electrolyte, leading to a consistent current density distribution. The rate at which reactants reach the electrode can be controlled by varying the rotation speed. At high enough rotation speeds, the reaction becomes mass-transport limited, meaning the rate of reactant supply solely determines the current. The RoDSE technique has been used to fabricate bulk batches of carbon-supported metal nanoparticle catalysts with high loading and uniform dispersion. For example, it has been employed to deposit gold and platinum nanoparticles onto Vulcan carbon, carbon nano-onions (CNOs), carbon nanotubes, and graphene oxide supports in a controlled manner. The key innovation is that the support material (carbon, etc.) is integrated into the deposition process, yielding a catalyst powder ready for use. RoDSE offers several advantages in catalyst preparation that directly benefit the performance of oxygen reduction reaction (ORR) catalysts. First, it enables precise control over nanoparticle size, distribution, and loading. Because the deposition is electrochemical, factors such as deposition potential, time, and metal-ion concentration can influence the nucleation and growth of particles. Researchers have shown that RoDSE can produce nanoparticles in the few-nanometre size range (e.g., ~4 nm Pt on carbon [21]) with a narrow size distribution. This is an optimal size range for ORR catalysts: small enough to maximise surface area and Pt atom utilisation, but not so small that most atoms are under-coordinated (which can reduce specific activity or cause dissolution). Additionally, RoDSE tends to deposit metal directly onto the support surface without the use of capping agents or surfactants. In the chemical colloidal synthesis of nanoparticles, stabilisers are often required to control growth, but those can poison catalyst surfaces. RoDSE, being electrochemical, can nucleate particles on the support without the need for such additives, yielding clean surface sites for the ORR. Another benefit is high dispersion and avoidance of agglomeration. Since the support particles are well separated in the slurry during deposition, the metal nanoparticles nucleate on those support surfaces rather than clumping together. This produces a catalyst where each nanoparticle is anchored on the support, preventing them from aggregating into larger, less active clusters. The result is a high utilisation of the metal: more of it is exposed to ORR participation. Moreover, RoDSE can create bulk quantities of catalyst in a single batch, which is vital for scale-up. Traditional RDE-based deposition is limited by the geometric area of the electrode (only a small amount of catalyst can be deposited at a time). Still, RoDSE effectively turns the entire electrolyte volume into a reactor for catalyst powder production. This ability to produce tens of milligrammes (or more) of catalyst in one process is valuable for research and potentially for industry, as it aligns with mass production needs.

The RoDSE method enables surface engineering and composite formation to enhance oxygen reduction reaction (ORR) catalysts. For instance, it is possible to co-deposit multiple metals or insert a second step, such as spontaneous galvanic displacement (SGD), to create core-shell structures [20]. Similarly, RoDSE can occlude nanoparticles within conductive oxide matrices (e.g., Pt deposition in a Vulcan-CeO_2 slurry to make a Pt/CeO_2/C catalyst [22]). Such hybrid structures can offer improved durability (oxide supports can mitigate carbon corrosion and provide synergistic catalytic sites). By controlling the electrodeposition environment, RoDSE also allows fine-tuning of metal loading—one can deposit just enough metal to create isolated single atoms or clusters on carbon (potentially useful for single-atom catalysts (SACs)) or continue deposition to get a higher loading of nanoparticles for maximum activity. This flexibility makes RoDSE a powerful tool for exploring various catalyst architectures (single atoms, clusters, core-shell NPs, etc.) relatively straightforwardly. Compared to conventional catalyst synthesis methods (like wet chemical reduction, impregnation, or thermal decomposition), RoDSE offers better integration of active material with the support and often superior purity. Traditional methods often require post-synthesis steps to attach nanoparticles to supports or remove stabilisers; RoDSE accomplishes this in one step, electroplating the active metal exactly where it needs to be—on the support. It also operates under mild conditions (room temperature, aqueous or glycol-based electrolytes), which can be gentler on supports and allow the use of supports that might not survive high-temperature treatments. Furthermore, RoDSE inherently yields a ready-to-use catalyst: the product is a supported catalyst powder that can be filtered, washed, and directly applied to an electrode. The avoidance of surfactants means there are no surface contaminants that might block catalytic sites. And because an applied potential drives the deposition, RoDSE can produce metastable alloy compositions or unusual structures that may be difficult to obtain *via* thermal methods. For example, one could electrodeposit a Pt-transition metal alloy [20] onto carbon at room temperature, even if that alloy might not form spontaneously by simple chemical reduction. The rapid electrochemical nucleation can "freeze in" specific compositions. Lastly, RoDSE's scalability and reproducibility are attractive for research and development: it is relatively easy to control the process parameters and to produce consistent batches of catalyst, which is essential when screening ORR materials for commercialisation. In summary, RoDSE is a valuable addition to the toolkit of ORR catalyst development, enabling the synthesis of high-performance catalysts (including PGM and PGM-free types) with controlled nanostructures and improved dispersion.

Figure 2 illustrates the core components and mechanism of the rotating disc slurry electrodeposition (RoDSE) cell. On the left is the three-electrode arrangement: a glassy-carbon rotating disc electrode (RDE) serves as the working electrode; a reversible hydrogen electrode (RHE) provides a stable reference potential, and a platinum-wire counter electrode completes the circuit. The enlarged cutaway on the right shows how, as the RDE spins, convective currents (red arrows) drive metal-ion precursors (cyan) through the carbon-particle slurry (black) towards the electrode surface. There, at a controlled applied potential, M^{n+} ions are reduced to M^0 and nucleate as metal nanoparticles (orange), which immediately adhere to the suspended carbon support, forming uniform metal/carbon hybrid catalysts. This configuration decouples the support material from a fixed substrate, ensuring homogeneous nanoparticle growth throughout the entire volume of the slurry.

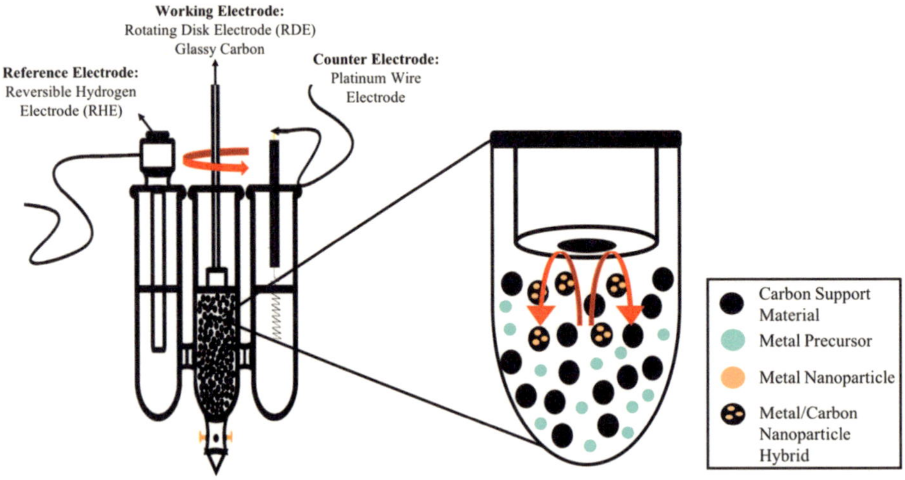

Figure 2.
General setup of the RoDSE technique [19]. Electrochemical arrangement of a three-electrode cell using a reversible hydrogen electrode (RHE) as the reference electrode, glassy carbon as the rotating working electrode, and a platinum coiled wire as the counter electrode in 0.1 M H_2SO_4. At the centre of the cell, precursor solution aliquots are added to a carbon slurry. (Reproduced with permission of the publisher).

The RoDSE method (**Figures 1** and **2**) will likely contribute to the future development of electrocatalyst synthesis. Its ability to produce bulk quantities of well-dispersed catalyst will be valuable for scaling up the most promising catalysts from lab scale to prototype devices. For non-PGM catalysts in particular, RoDSE could enable the mass manufacture of Fe–N–C [23] or transition-metal oxide catalysts with controlled nanostructures. For example, one could envision using RoDSE to deposit trace amounts of iron onto a nitrogen-doped carbon support in a single step, thereby creating an Fe–N–C catalyst without the high-temperature pyrolysis typically required. This might yield a more uniform distribution of active or novel active site environments. Such hybrid catalysts might harness multiple ORR pathways and improve overall activity or durability. The comparative advantage of RoDSE in producing uniform and agglomeration-free nanoparticles aligns well with the needs of next-generation ORR catalysts, where maintaining a high density of isolated active sites is key.

3. Materials synthesised using RoDSE

3.1 Pt-based catalysts

As a versatile and energy-efficient electrochemical technique for synthesising electrocatalysts, RoDSE has been used to synthesise a wide range of materials. RoDSE has been extensively utilised for the synthesis of platinum nanoparticles on various carbon supports, demonstrating exceptional control over particle size and distribution. The technique enables the preparation of Pt/Vulcan XC-72R catalysts with well-dispersed nanoparticles [21], showcasing superior electrochemical performance compared to commercial catalysts. Platinum (Pt) and its alloys [20] have long been considered the gold standard for catalysis [24] due to their high activity and stability. In alkaline media, Pt-based catalysts continue to demonstrate excellent performance. Pure Pt catalysts [25] serve as a benchmark for ORR activity in alkaline conditions. They typically follow a four-electron pathway, directly reducing oxygen to hydroxide ions without forming peroxide intermediates. The high activity of Pt is attributed to its optimal binding energy with oxygen species, allowing for efficient adsorption and desorption of reactants and products.

Alloying Pt with other metals (Ni, Co, Fe, etc.) can *dramatically* improve ORR activity [20] by tuning Pt's electronic structure and surface geometry. A landmark example is the Pt_3Ni (111) single-crystal surface [26], which exhibits an order of magnitude higher ORR activity than pure Pt (111). Similarly, nanoparticle alloys like Pt-Ni or Pt–Co show enhanced specific activity due to optimised oxygen binding energy (d-band centre shifts) and favourable ensemble effects. **Figure 3** panels (a–i) collectively illustrate the size, distribution, and morphology of Pt nanoparticles electrodeposited by RoDSE on carbon supports. In each row, the leftmost image (scale bar = 20 nm) shows low-magnification transmission electron microscopy (TEM) highlighting the uniform "speckling" of dark Pt particles across the lighter carbon substrate, while the centre image (scale bar = 10 nm) zooms in on individual clusters, revealing mostly spherical to slightly faceted nanoparticles with clear lattice fringes.

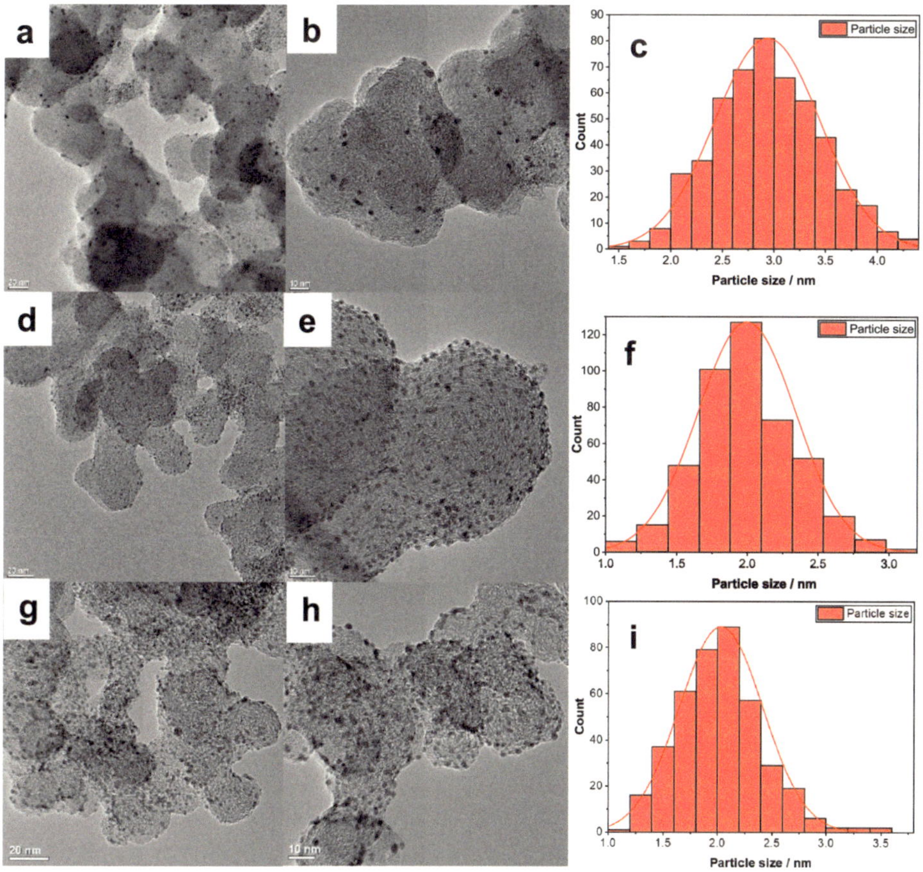

Figure 3.
Transmission electron microscopy images [20] of PtM/CBV electrocatalysts and their respective particle size distributions: (a–c) PtNi/CBV, (d–f) PtCo/CBV, and (g–i) PtCu/CBV. (Reproduced with permission of the publisher).

The rightmost histograms quantify these observations: for Pt (panel c), particle diameters span roughly 1.5–4.0 nm with a mean near 3.0 nm; on (panel f), a tighter distribution (1.2–3.0 nm) centres around 2.1 nm; and on (panel i), particles range from about 1.0–3.5 nm with a mean of 2.3 nm.

One of the most significant achievements in RoDSE synthesis involves platinum electrodeposition on carbon nano-onions [17] (CNOs), also known as multilayer fullerenes. X-ray photoelectron spectroscopy analysis reveals that 37% of the platinum on CNOs exists as metallic Pt. In comparison, 63% exhibits higher binding energies indicative of higher oxidation states or the presence of Pt atoms and clusters. The resulting Pt/CNO materials demonstrate excellent thermal stability and enhanced electrochemical performance compared to commercial Pt/Vulcan catalysts.

RoDSE enables the decoration of single-walled carbon nanotubes (SWCNTs) [27] with platinum nanoparticles for specialised sensing applications. The electrodeposited nanoparticles achieve an average particle size of 4.1 ± 0.8 nm, and the resulting Pt/SWCNT composites demonstrate improved methane sensing properties in the concentration range of 50–200 ppm at room temperature. The rotating disc slurry electrodeposition (RoDSE) methodology, combined with spontaneous galvanic displacement (SGD) [20], has successfully synthesised a series of transition metal-platinum alloy catalysts with remarkably low platinum loadings and exceptional electrochemical performance. The synthesis process involves a two-step approach where first-row transition metals (nickel, cobalt, and copper) are electrodeposited onto carbon black Vulcan XC-72R support using applied potentials of −0.75 V versus RHE for nickel and cobalt and − 0.80 V versus RHE for copper in 0.1 M $KClO_4$ electrolyte (**Figure 4**). These transition-metal nanoparticles are subsequently modified with platinum precursor *via* spontaneous galvanic displacement using 10 mM H_2PtCl_6 solution, resulting in PtNi/CBV, PtCo/CBV, and PtCu/CBV catalysts with ultra-low platinum loadings of approximately 1.7 ± 0.2%, 1.9 ± 0.3%, and 3.0 ± 0.1%, respectively (**Figure 3**).

RoDSE also facilitates synthesising platinum on Y-zeolite/carbon Vulcan XC-72R composites, showcasing the technique's flexibility in forming hierarchical catalyst structures. The Pt/Y/V catalyst, containing 14 wt% Pt, demonstrates greater activity than commercial Pt/V catalysts with 20 wt% Pt for methanol electro-oxidation in alkaline environments. This improved performance stems from distinctive nucleation and growth patterns within the zeolite framework, leading to both nanoparticles and tetrahedral shapes. Additionally, RoDSE enables the synthesis of intricate composite materials, such as platinum/cerium oxide systems, through occlusion electrodeposition methods.

The Pt/CeO_{2-x}/Vulcan XC-72R [22] catalysts demonstrate exceptional activity for methanol and ethanol oxidation reactions in alkaline conditions, with mass activity increases of 2.0× and 2.4× compared to commercial catalysts. The unique aspect of this synthesis involves the growth of platinum nanoparticles within cerium oxide through an occlusion electrodeposition mechanism. The electrochemical synthesis of polyaniline on onion-like carbon nanoparticles using RoDSE represents an innovative approach to creating conductive polymer-carbon hybrids. These particles are characterised using Raman and Fourier transform infrared spectroscopy, as well as scanning electron microscopy. The combination of carbon compounds and conductive polymers results in new materials with potential practical applications. The ultra-small nanostructure capabilities of RoDSE suggest potential applications in quantum dot (QD) synthesis beyond the current iron-based systems. Future developments may include the synthesis of semiconductor quantum dots, particularly those

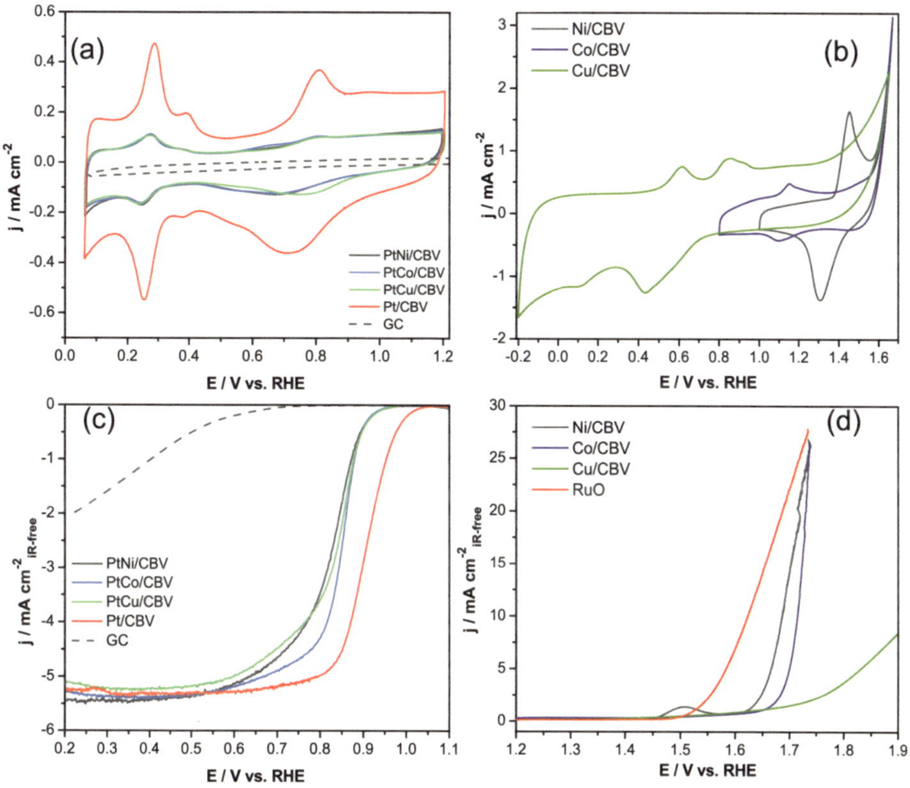

Figure 4.
Cyclic voltammograms [20] of (a) PtNi/CBV, PtCo/CBV, PtCu/CBV, and commercial Pt/CBV and (b) Ni/CBV, Co/CBV, and Cu/CBV at their respective potential windows at 25 mV/s. (c) ORR polarisation curves of PtNi/CBV, PtCo/CBV, PtCu/CBV, and commercial Pt/CBV at 10 mV/s at 1600 rpm. (d) Oxygen evolution reaction (OER) polarisation curves of Ni/CBV, Co/CBV, Cu/CBV, and commercial RuO at 10 mV/s and 1600 rpm. All the experiments were done in 0.1 M KOH. (Reproduced with permission of the publisher).

incorporating multiple elements to achieve tunable optical and electronic properties. The precise control over particle size achieved through RoDSE makes it suitable for quantum confinement applications.

3.2 Non-platinum-based electrocatalysts

Replacing Pt with earth-abundant metals like iron, cobalt, or nickel (and their alloys) is appealing due to cost. Alloy combinations such as Fe–Co [10] or Fe–Ni [28] have been explored as ORR catalysts, especially in alkaline media with higher stability. These non-PGM alloys can catalyse ORR by providing active metallic sites for O_2 adsorption, but in practice, they often require protective matrices to prevent corrosion. Silver-palladium [19] Bimetallic nano catalysts synthesised *via* RoDSE (**Figure 5**) represent a cost-effective approach to improving oxygen reduction reaction performance in alkaline media. The technique enables three different electrodeposition strategies: alternating, sequential, and simultaneous Ag and Pd electrodeposition on Vulcan XC-72R. Among these approaches, the alternated method yields (**Figure 6**) the best performance in terms of onset potential, peak current density, and electron transfer characteristics in O_2-saturated 0.1 M KOH.

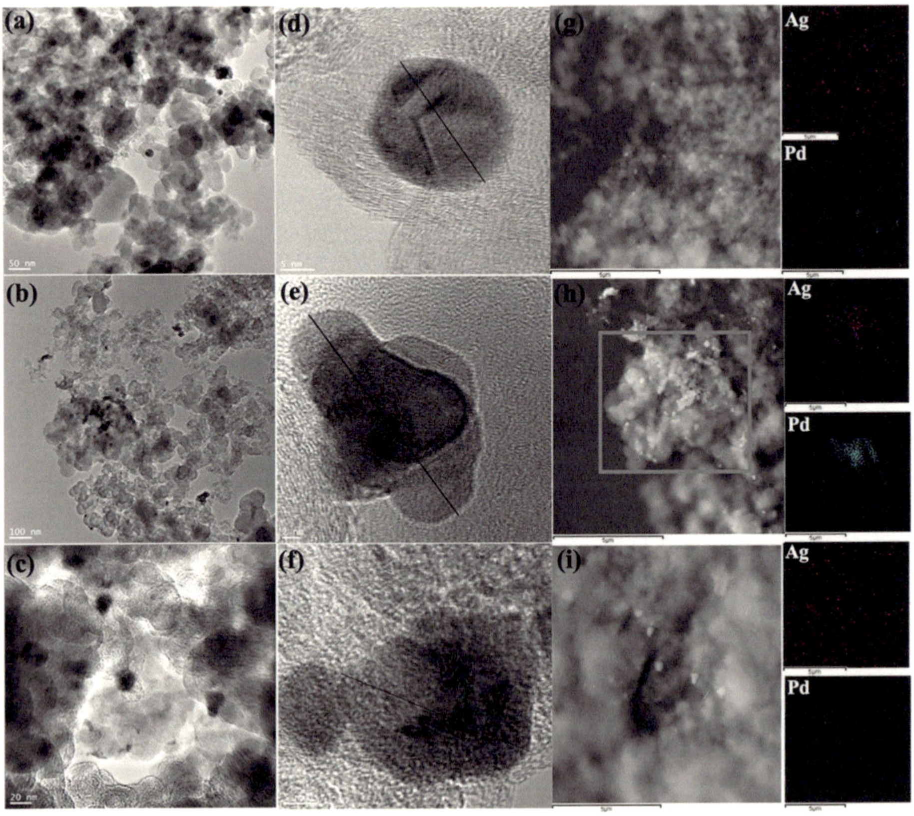

Figure 5.
HRTEM images [19] of Ag/Pd/Vulcan catalyst prepared by RoDSE alternated (a and d), sequential (b and e), and simultaneous (c and f) methods; STEM for elemental mapping of Ag and Pd in (g) alternated, (h) sequential, and (i) simultaneous methods. (Reproduced with permission of the publisher).

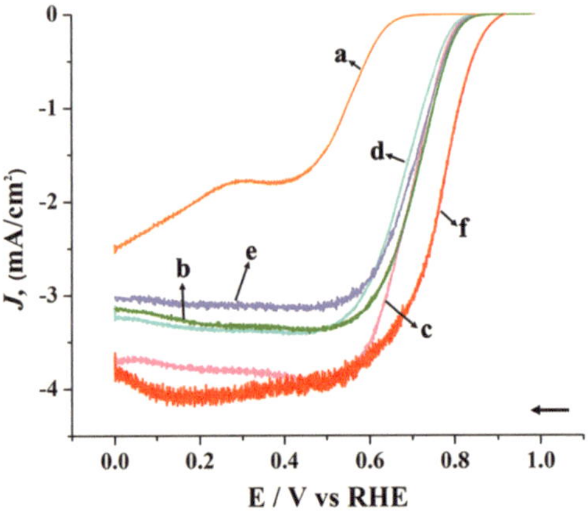

Figure 6.
Linear sweep voltammograms [19] from RDE measurement in O_2-saturated 0.1 M KOH solution at a scan rate of 20 $mV\ s^{-1}$ at 1600 rpm for the following catalysts: (a) Ag/Vulcan, (b) Pd/Vulcan, (c) Ag/Pd/Vulcan alternated, (d) Ag/Pd/Vulcan sequential, (e) Ag/Pd/Vulcan simultaneous, and (f) 20% Pt/Carbon Vulcan. The arrows indicate the potential sweep direction, from right to left to negative values. (Reproduced with permission of the publisher).

The resulting Ag-Pd bimetallic catalysts exhibit face-centred cubic crystal facets with approximate particle diameters of 23 nm. As in **Figure 5**, all the electrocatalysts show the presence of both Pd and Ag metals. These materials demonstrate enhanced ORR activity compared to their monometallic counterparts, highlighting the synergistic effects achieved through controlled bimetallic synthesis.

The synthesis of trimetallic nickel-cobalt-iron nanoparticles [29] using RoDSE represents a significant breakthrough in the development of platinum-group-metal-free electrocatalysts. The electrodeposition process utilises negative potentials (−1.05 V vs. Ag/AgCl-sat KCl) in saturated argon environments, with metal salt solution aliquots added every 20 minutes over a total of 12 additions.

The resulting Ni-Co-Fe trimetallic catalysts demonstrate comparable mass activity and durability to commercial 20% Pt/Vulcan, with an onset potential of 0.78 V vs. RHE and limiting current density of 5.56 mA/cm^2. X-ray diffraction studies and transmission electron microscopy confirm the formation of well-dispersed trimetallic nanoparticles on the carbon support.

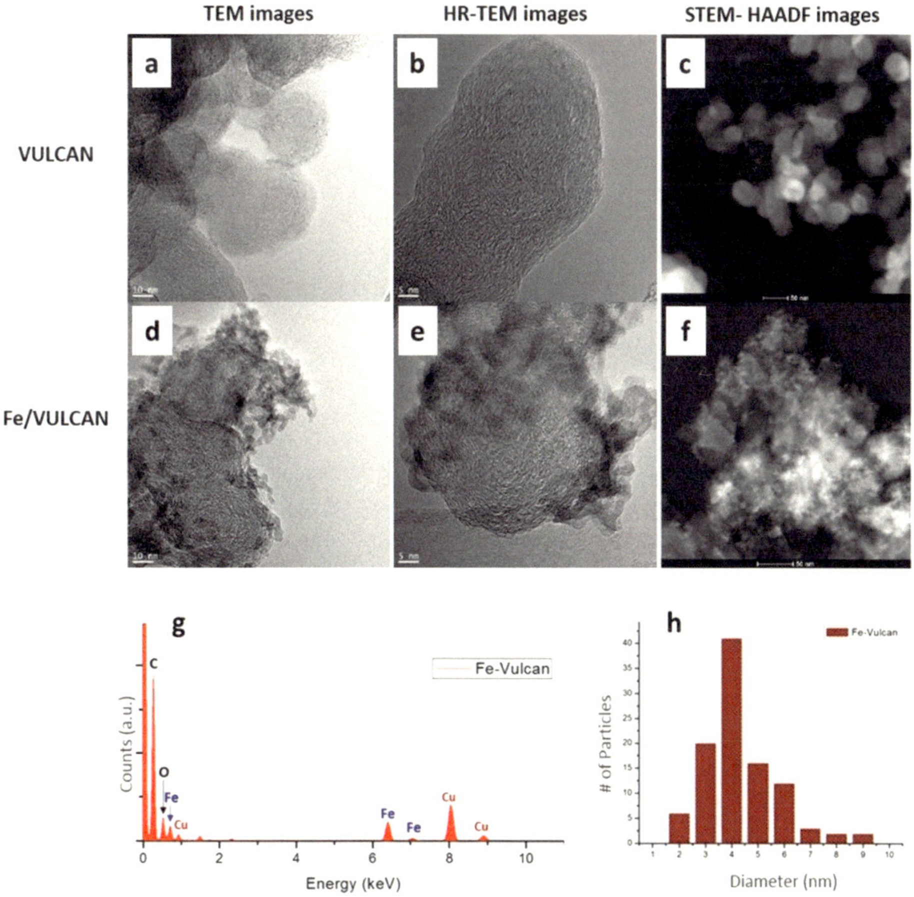

Figure 7.
Transmission electron microscopy (TEM) [5] (a,d), high-resolution transmission electron microscopy (HR-TEM) (b,e), and high-angle annular dark-field scanning transmission electron microscopy (HAADF-STEM) (Contrast-Z) (c,f), of Vulcan and the Fe/Vulcan catalyst electrodeposited by the RoDSE technique. The energy dispersive X-ray spectroscopy (EDS) analysis evidences the presence of iron (g). The particle size histogram shows a narrow size distribution of about 4 nm of Fe-based QDs onto Vulcan (h). (Reproduced with permission of the publisher).

3.3 Advanced quantum dot systems

In advanced quantum dot (QD) systems, electrocatalysis benefits from quantum confinement: the sub-10 nm size of QDs creates discrete energy levels and tunable band gaps that can be aligned precisely with the redox potentials of targeted reactions such as the oxygen reduction, hydrogen evolution, or CO_2 reduction reactions. The high surface-to-volume ratio exposes a dense population of under-coordinated atoms—intrinsically more reactive sites—while engineered core-shell or alloyed QD structures enhance charge separation and suppress recombination. When these QDs are integrated onto conductive supports *via* techniques like RoDSE, they form robust, high-surface-area electrodes that combine rapid electron transport with size-tunable catalytic activity, enabling next-generation electrocatalytic devices with greater efficiency and selectivity than traditional nanoparticle catalysts.

The ultra-small nanostructure capabilities of RoDSE suggest potential applications in quantum dot synthesis [5] that extend beyond current iron-based systems. Future developments may include the synthesis of semiconductor quantum dots, particularly those incorporating multiple elements to achieve tunable optical and electronic properties. The precise control over particle size achieved through RoDSE makes it suitable for quantum confinement applications.

The ultra-small nanostructure capabilities of RoDSE suggest potential applications in quantum dot synthesis that extend beyond current iron-based systems. As shown in **Figure 7**, the average size distribution of the iron quantum dots is around four nanometres, with an even distribution. Future developments may include the synthesis of semiconductor quantum dots, particularly those incorporating multiple elements to achieve tunable optical and electronic properties. The precise control over particle size achieved through RoDSE makes it suitable for quantum confinement applications.

4. Future material predictions

The rotating disc slurry electrodeposition (RoDSE) technique presents a significant promise for synthesising a diverse range of advanced functional materials. Building on its proven versatility in fabricating finely dispersed metallic nanoparticles, RoDSE's scope is rapidly expanding to address novel, sophisticated material platforms. Based on current progress and emerging trends in materials science, several key directions for future RoDSE-mediated synthesis can be anticipated.

4.1 High-entropy transition-metal carbides, nitrides, and chalcogenides

The remarkable success of RoDSE in generating high-entropy alloys (HEAs) through the simultaneous electrodeposition of multiple transition-metal species strongly suggests a pathway for extending this methodology to synthesising high-entropy metal carbides, nitrides, and chalcogenides. By replacing typical aqueous solutions of transition-metal salts with solutions containing cyanamide, thiourea, thiocyanates, or related nitrogen/sulphur/carbon donor ligands, RoDSE could deposit amorphous or nanocrystalline precursors at cathodic potentials typically ranging from −0.8 to −1.2 V vs. RHE. Subsequent thermal annealing under controlled atmospheres (methane, ammonia, or hydrogen sulfide gas, respectively) would crystallise these precursors into nanometre-sized high-entropy transition-metal carbides

(M_xC_y), nitrides (M_xN_y), or chalcogenides (M_xS_y), each with distinct electronic and catalytic properties. The precise hydrodynamic control inherent in RoDSE would ensure the synthesis of ultrafine particles (typically below 5 nm) with homogenous chemical distributions, enabling their direct integration into high-performance electrocatalysts or protective coatings. Furthermore, by leveraging combinatorial methodologies inherent in RoDSE—rapid variation of bath composition and electrodeposition potential—vast libraries of materials could be rapidly generated, facilitating high-throughput identification of compositions exhibiting superior electrocatalytic activities or stability.

4.2 MXene-integrated metal/oxide heterostructures

MXenes [30], two-dimensional transition-metal carbides or nitrides, have attracted substantial attention due to their exceptional conductivity and catalytic capabilities. However, current MXene-based heterostructures suffer from limited interlayer stability and aggregation issues. RoDSE could be strategically employed after the post-electrophoretic deposition (EPD) of MXene nanosheets to directly intercalate precisely sized metallic or oxide nanoparticles within the MXene layers. By maintaining hydrodynamic conditions through rotation speeds of 1000–1500 rpm, RoDSE would prevent MXene layer restacking and aggregation, creating ordered, hierarchical, nanoparticle-pillared MXene frameworks. For instance, the targeted deposition of catalytic nanoparticles, such as Ni-Co-P or $CoMn_2O_4$ [31], could significantly enhance electrochemical performance, particularly in applications like lithium-sulphur or zinc-air batteries (ZABs), where controlled interlayer spacing and robust catalytic activity are crucial.

4.3 Layered double hydroxide (LDH) core-shell catalysts

An exciting avenue for RoDSE advancement involves synthesising sophisticated core-shell catalysts that combine metallic nanoparticle cores with outer shells of layered double hydroxides (LDHs) [32]. LDHs exhibit unique catalytic and redox properties but typically require multistep synthetic procedures. Utilising pulse-reversal RoDSE sequences—cathodic pulses (+0.3 to +0.5 V vs. RHE) for hydroxide shell formation—complex core-shell structures such as Pt-alloy/LDH hybrids could be synthesised directly. Such RoDSE-derived core-shell catalysts would offer tailored bifunctionality, where the conductive metallic core enhances electron transport, and the LDH shell provides active sites for catalytic redox reactions (e.g., oxygen evolution and reduction). The uniformity and thickness control of LDH shells achievable through RoDSE promise superior durability, enhanced activity, and resistance against degradation typically seen in traditional catalyst materials.

4.4 Two-dimensional MOF-templated single-atom arrays

Integrating single-atom catalysts (SACs) with metal-organic framework (MOF) [33]-derived carbons presents a vital strategy to maximise catalytic efficiency through atomic dispersion of metals. RoDSE's capacity for atomic-level deposition precision could transform synthesis strategies for SACs. Specifically, metal ions such as Fe, Co, or Pt could be deposited as isolated atoms directly onto the surface of pre-synthesised two-dimensional MOF nanosheets dispersed in a slurry under well-controlled potentials (typically near their respective underpotential deposition (UPD) windows). A controlled post-treatment pyrolysis at moderate temperatures (~700–800°C) would convert the MOF nanosheets into nitrogen-rich graphene analogues, securely anchoring the isolated metal atoms as active metal-N_4 sites. Such a strategy could achieve unprecedentedly high loadings of catalytically active single-atom sites, providing extraordinary performance in key electrocatalytic applications such as oxygen reduction reactions in fuel cells or carbon dioxide reduction.

4.5 Perovskite-metal nanocomposites

The integration of metallic nanoparticles into perovskite-based frameworks is gaining attention, especially for high-temperature catalytic and electrocatalytic applications such as solid oxide fuel cells. RoDSE could facilitate the direct incorporation of metallic nanoparticles (e.g., Ag, NiFe alloys) into porous perovskite scaffolds (e.g., $Ba_{0.5}Sr_{0.5}Co_{0.8}Fe_{0.2}O_{3-\delta}$ or $CaTiO_3$) [34]. Sequential RoDSE deposition cycles, first cathodically depositing metallic nanoparticles within the pores of the perovskite frameworks, followed by controlled anodic pulses to regenerate oxygen vacancies or repair lattice defects, could yield stable and highly active perovskite-metal composite electrocatalysts. The resulting synergistic interactions between metal nanoparticles and perovskite lattice structures could lead to substantial improvements in electrocatalytic performance, durability, and ionic conductivity.

5. Applications of ORR

Because ORR is the cathodic reaction in many electrochemical energy devices, improvements in ORR catalysis have a far-reaching impact. Here, we briefly survey the major applications of ORR and the specific demands they impose on catalysts:

5.1 Proton-exchange membrane fuel cells (PEMFCs)

In PEM fuel cells (which have an acidic polymer electrolyte, e.g., Nafion), hydrogen is oxidised at the anode and oxygen is reduced at the cathode. ORR in PEMFCs is a $4e^-$ process to form water, and it is well known to be the bottleneck for PEMFC efficiency. State-of-the-art PEMFCs use Pt-based catalysts on the cathode (typically Pt/C) to accelerate ORR. Even so, a significant overpotential (hundreds of millivolts) is required to drive practical currents, meaning precious energy is lost as heat due to slow ORR kinetics. This limits the fuel cell's voltage output and efficiency. For example, while the thermodynamic cell voltage is ~1.23 V, a PEMFC under load may deliver only ~0.6–0.7 V per cell, much of the loss attributable to the cathode [35]. The challenge in PEMFC ORR catalysis is finding catalysts that can approach the activity of Pt but at a lower cost and withstand the acidic environment. Non-PGM catalysts have struggled in this context because the acidic medium corrodes most base metals and leaches active sites. Thus, PEMFCs remain reliant on PGM catalysts, and research is focused on PGM alloy catalysts and improved electrode structures to boost performance and reduce Pt loading. Additionally, there is interest in direct methanol fuel cells (DMFCs) [36] and other fuel cells where ORR must tolerate fuels or impurities; here, Pt-based ORR catalysts face poisoning by crossover methanol or CO, so more poison-tolerant catalysts (or bifunctional catalysts that can also oxidise the crossover fuel) are sought.

5.2 Alkaline fuel cells (AFCs) and anion-exchange membrane fuel cells (AEMFCs)

Alkaline fuel cells, including those with a liquid KOH electrolyte and the newer solid anion-exchange membrane fuel cells (AEMFCs), also rely on ORR at the cathode (producing OH^- in alkaline media). The alkaline environment allows a broader range of catalysts to be used, because many transition metals and oxides that would dissolve in acid are stable as ORR catalysts in base. Historically, AFCs have also used Pt catalysts (the Apollo space programme's alkaline fuel cells, for instance, used platinum electrodes). Still, one can substitute or augment Pt with silver and metal oxides in alkaline conditions. Modern AEMFC research has shown particular interest in applying Fe–N–C [37] or Co–N–C non-precious catalysts at the cathode, leveraging their good performance in alkaline environments. The best Fe–N–C catalysts in AEMFCs can now approach the performance of Pt catalysts, at least initially, although durability and integration with alkaline ionomers remain challenges. AEMFCs promise the efficiency of PEMFCs but without precious metals, which would be a game-changer for fuel cell economics. To achieve this, the ORR catalyst in AEM systems must be optimised for the high-pH environment and tailored to work with the anion-exchange membrane (which can have different constraints than a liquid electrolyte). Overall, ORR research for alkaline fuel cells enables the possibility of entirely PGM-free fuel cell stacks.

5.3 Metal-air batteries (Zn-air, Li-air, etc.)

In metal-air batteries [3], metal anodes (such as zinc or lithium) are coupled with an air-breathing cathode where ORR occurs during the discharge (and the reverse oxygen evolution reaction, OER, occurs during charging for rechargeable systems). ORR is a vital reaction in these batteries, enabling the cell to capture energy from oxygen in the air. A primary Zn-air battery [3] (used once until the zinc is consumed) relies on ORR to generate electricity from zinc oxidation; a rechargeable Zn-air (or Li-air) needs a bifunctional catalyst to do ORR during discharge and OER during charge. The ORR catalysts used in metal-air systems are often similar to those in fuel cells: Pt or other noble metals can catalyse ORR, but due to cost, metal-air developers favour non-precious catalysts like manganese oxides, perovskites, or carbon-based catalysts. For instance, in Zn-air batteries (ZABs), manganese dioxide and related oxides have been widely used as ORR catalysts because of their favourable activity and low cost. ORR in ZABs occurs in concentrated alkaline electrolyte (e.g., KOH), where MnO_2 and Co_3O_4 catalysts are stable and effective. As noted earlier, Pt is an excellent ORR catalyst. Still, in a strong KOH environment, it can be partially hindered by OH adsorption and offers no OER capability; thus, bifunctional air electrodes often combine a suitable ORR catalyst (like Pt or MnO_2) with a suitable OER catalyst (like IrO_2 or NiFe-layered hydroxide). The need for bifunctionality is a unique aspect of rechargeable metal-air batteries. In the case of Li-air batteries (also known as lithium-oxygen batteries), the ORR occurs in a non-aqueous electrolyte and produces lithium peroxide (Li_2O_2) or related species in the cathode. The ORR in Li-air cells [23] is notoriously tricky; the kinetics are even slower, and the reaction product (Li_2O_2) can clog the porous cathode. One of the significant drawbacks of Li-air batteries is indeed the sluggish oxygen reduction reaction (ORR) at the cathode, which limits both the discharge rate and capacity. Considerable research is being devoted to finding catalysts (often based on bifunctional carbon/oxide composites or redox mediators) that can facilitate the oxygen reduction reaction (ORR) in the non-aqueous environment of Li-air cells. In summary, whether for Zn-air or Li-air, the performance of metal-air batteries is tightly linked to

ORR catalyst performance: a better ORR catalyst yields higher power and efficiency. Thus, advances in ORR catalyst materials (especially inexpensive ones) directly benefit the development of metal-air energy storage for applications like hearing aids, electric vehicle batteries (in the case of Li-air, in the future), and grid storage.

6. Conclusion and outlook

In summary, a broad spectrum of materials is under active investigation to catalyse the ORR in fuel cells and metal-air batteries, each with its strengths and weaknesses. PGM alloys (like Pt-Ni and Pt-Co) remain the benchmark for activity, particularly in acidic fuel cells, and innovations in alloying and nanostructuring are helping reduce precious metal usage and improve stability. Non-PGM alloys (Fe, Co, and Ni-based) offer a low-cost alternative, demonstrating viable performance in alkaline systems and benefiting from strategies such as heteroatom doping and encapsulation to enhance their activity and corrosion resistance. Spinel oxides [38] provide a robust option for metal-air battery cathodes and alkaline fuel cells, leveraging multivalent cation chemistry to achieve high ORR rates; ongoing research into their composition and structure (e.g., mixed-metal spinels) is yielding catalysts that approach noble-metal behaviour in basic media. The broad class of transition metal M–N–C catalysts [39] has demonstrated that with the correct atomic structure (Fe/Co coordinated by N and other heteroatoms in carbon), one can attain remarkable ORR performance without precious metals—the focus now is on overcoming durability challenges and bridging the remaining activity gap in acidic environments. The advent of single-atom catalysts [39] pushes these ideas further, atomically engineering the active sites to maximise efficiency and tailor reaction pathways; continued advances in synthesis and support design will determine how far SACs can go in replacing bulk Pt in real devices.

Future ORR catalyst development will likely be an integrative process. We may see hybrid cathodes that combine, for example, a small amount of Pt alloy or single-atom sites to handle high-current conditions with a bulk of doped carbon or oxide catalyst to reduce cost and enhance stability (creating a composite that capitalises on each component's advantages). Moreover, insights from each category are beginning to converge. For instance, knowledge of single-atom active site structure can inform the design of active sites in carbon or alloy catalysts, and vice versa. In fuel cells and metal-air batteries, there is a drive for activity, stability, and bifunctionality (especially for rechargeable metal-air batteries [3] where ORR and OER occur on the same electrode). This might spur research into catalysts that can switch oxidation states reversibly (as some spinels and perovskites can) or pair oxygen reduction reaction (ORR) catalysts with oxygen evolution reaction (OER) catalysts.

From a commercialisation standpoint, the outlook is cautiously optimistic. In the near term, improved PGM catalysts (Pt alloys, nanostructured Pt, etc.) will continue to be used in applications like PEM fuel cells—the focus there is on reducing PGM loading to reduce cost (for instance, automakers are targeting $<0.125\ mg_{Pt}/cm^2$ total loading). In parallel, AEM fuel cells are emerging as a platform where non-PGM ORR catalysts can be realistically used, as the alkaline environment allows Fe/Co-based catalysts to operate without dissolution. We may soon see AEMFC prototypes with PGM-free cathodes if durability can be proven. For metal-air batteries, non-PGM bifunctional catalysts (e.g., based on perovskite oxides or doped carbons) are already enabling decent cycle life in Zn-air cells at lab scale; the challenge is to maintain performance over many charge-discharge cycles, where catalyst stability and air electrode carbonation (CO_2 intake) are concerns. Continued research into protective catalyst layer designs and efficient utilisation will be necessary.

In conclusion, developing ORR catalysts for alkaline electroreduction is a vibrant field at the intersection of materials science, electrochemistry, and engineering. PGM catalysts like platinum have set a high benchmark for activity and stability, but their cost drives the search for alternatives. Non-PGM catalysts, once far behind, have rapidly advanced, particularly in alkaline media, and some now rival precious metals in performance. The RoDSE method and other innovative synthesis techniques accelerate this progress by enabling precise control over catalyst structure and composition at the nanoscale. As we advance, the combination of rational catalyst design (guided by understanding of ORR mechanisms and computational modelling) and advanced fabrication methods (like RoDSE, atomic layer deposition, etc.) will likely yield a new generation of ORR catalysts that fulfil the dream of efficient, stable, and affordable energy conversion. Such catalysts are crucial for the broader adoption of fuel cells and metal-air batteries, key technologies for a sustainable energy future. The advances reviewed here underscore a clear trajectory: improving ORR catalysis from the atomic to the device level, with the ultimate goal of commercial viability and impactful clean energy applications.

Acknowledgements

This work is supported as part of the Center for Alkaline-Based Energy Solutions (CABES), an Energy Frontier Research Center funded by the U.S. Department of Energy (DOE), Office of Science, Basic Energy Sciences (BES), under Award # DE-SC0019445. XLEAP, NSF-BIO 2330043, supported a portion of this effort. Chatterjee acknowledges Sebastián Plaza Sierra for helping with citations. CRC acknowledges the STARs Award (2021) from the University of Texas System and the Ralph and Kathleen Ponce de Leon Professorship in Chemistry.

Author details

Suzatra Chatterjee[1*], Abayomi Omoogun[1], Ruma Rani Paul[1], Subhajit Bhunia[2] and Carlos R. Cabrera[1]

1 Department of Chemistry and Biochemistry, University of Texas at El Paso, El Paso, Texas, United States

2 Department of Chemistry and Biochemistry, University of Utha, Salt Lake City, UT, USA

*Address all correspondence to: schatterjee2@miners.utep.edu

References

[1] Eguizabal A, Uson L, Sebastian V, Hueso L, Pilar Pina M. Efficient and facile tuning of Vulcan XC72 with ultra-small Pt nanoparticles for electrocatalytic applications. RSC Advances. 2015;**5**(110):90691-90697. DOI: 10.1039/C5RA16698E

[2] Cano ZP, Banham D, Ye S, Hintennach A, Lu J, Fowler M, et al. Batteries and fuel cells for emerging electric vehicle markets. Nature Energy. 2018;**3**(4):279-289. DOI: 10.1038/s41560-018-0108-1

[3] Li Y, Lu J. Metal–air batteries: Will they be the future electrochemical energy storage device of choice? ACS Energy Letters. 2017;**2**(6):1370-1377. DOI: 10.1021/acsenergylett.7b00119

[4] Ramaswamy N, Tylus U, Jia Q, Mukerjee S. Activity descriptor identification for oxygen reduction on nonprecious electrocatalysts: Linking surface science to coordination chemistry. Journal of the American Chemical Society. 2013;**135**(41):15443-15449. DOI: 10.1021/ja405149m

[5] Peña-Duarte A, Vijapur SH, Hall TD, Hayes KL, Larios-Rodríguez E, Pilar-Albaladejo JD, et al. Iron quantum dots electro-assembling on vulcan XC-72R: Hydrogen peroxide generation for space applications. ACS Applied Materials & Interfaces. 2021;**13**(25):29585-29601. DOI: 10.1021/acsami.1c05649

[6] Ramaswamy N, Mukerjee S. Fundamental mechanistic understanding of electrocatalysis of oxygen reduction on Pt and non-Pt surfaces: acid versus alkaline media. Advances in Physical Chemistry. 2012;**2012**(1):491604. DOI: 10.1155/2012/491604

[7] Paulus UA, Schmidt TJ, Gasteiger HA, Behm RJ. Oxygen reduction on a high-surface area Pt/Vulcan carbon catalyst: A thin-film rotating ring-disk electrode study. Journal of Electroanalytical Chemistry. 2001;**495**(2):134-145. DOI: 10.1016/S0022-0728(00)00407-1

[8] Zhou Z, Zhang H-J, Feng X, Ma Z, Ma Z-F, Xue Y. Progress of Pt and iron-group transition metal alloy catalysts with high ORR activity for PEMFCs. Journal of Electroanalytical Chemistry. 2024;**959**:118165. DOI: 10.1016/j.jelechem.2024.118165

[9] Anjana J, Muthukrishnan A. Effect of local pH change on non-PGM catalysts—A potential-dependent mechanistic analysis of the oxygen reduction reaction. Catalysis Science & Technology. 2022;**12**(20):6246-6255. DOI: 10.1039/D2CY01099B

[10] Zhu M, Zhang C. FeCo nanoalloys encapsulated in pod-like N-doped carbon nanotubes as efficient oxygen reduction reaction electrocatalysts for zinc-air batteries. Journal of Alloys and Compounds. 2022;**921**:166122. DOI: 10.1016/j.jallcom.2022.166122

[11] Hu C, Dai L. Carbon-based metal-free catalysts for electrocatalysis beyond the ORR. Angewandte Chemie, International Edition. 2016;**55**(39):11736-11758. DOI: 10.1002/anie.201509982

[12] Bhunia S, Chatterjee S, Cabrera CR. Platinum nanoparticle synthesis in engineered organic nanoscale reactors for efficient oxygen electro-reduction in alkaline conditions. MRS Communications. 2024;**14**(6):1159-1166. DOI: 10.1557/s43579-024-00547-2

[13] Bhunia S, Lu L, Chatterjee S, Garaga M, Mayoral A, Head AR, et al.

Unraveling Dimensional Tuning: From 2D to 3D in Covalent Organic Frameworks for Enhanced 2e– Oxygen Reduction Reaction. ACS Omega. 2025;**10**(18). DOI: 10.1021/acsomega.5c00568

[14] Escobar Morales B, Gamboa SA, Pal U, Guardián R, Acosta D, Magaña C, et al. Synthesis and characterization of colloidal platinum nanoparticles for electrochemical applications. International Journal of Hydrogen Energy. 2010;**35**(9):4215-4221. DOI: 10.1016/j.ijhydene.2010.01.040

[15] Carpenter MK, Moylan TE, Kukreja RS, Atwan MH, Tessema MM. Solvothermal synthesis of platinum alloy nanoparticles for oxygen reduction electrocatalysis. Journal of the American Chemical Society. 2012;**134**(20):8535-8542. DOI: 10.1021/ja300756y

[16] Santiago D, Cruz-Quiñonez M, Tryk DA, Cabrera CR. Pt/C catalyst preparation using rotating disk-slurry electrode (RoDSE) technique. ECS Transactions. 2007;**3**(21):35. DOI: 10.1149/1.2721515

[17] Santiago D, Rodríguez-Calero GG, Palkar A, Barraza-Jimenez D, Galvan DH, Casillas G, et al. Platinum electrodeposition on unsupported carbon nano-onions. Langmuir: The ACS Journal of Surfaces and Colloids. 2012;**28**(49):17202-17210. DOI: 10.1021/la3031396

[18] Vega-Cartagena M, Rojas-Pérez A, Colón-Quintana GS, Blasini Pérez DA, Peña-Duarte A, Larios-Rodríguez E, et al. Potential dependent Ag nanoparticle electrodeposition on vulcan XC-72R carbon support for alkaline oxygen reduction reaction. Journal of Electroanalytical Chemistry. 2021;**891**:115242. DOI: 10.1016/j.jelechem.2021.115242

[19] Vega-Cartagena M, Flores-Vélez EM, Colón-Quintana GS, Blasini Pérez DA, De Jesús MA, Cabrera CR. Silver–palladium electrodeposition on unsupported vulcan XC-72R for oxygen reduction reaction in alkaline media. ACS Applied Energy Materials. 2019;**2**(7):4664-4673. DOI: 10.1021/acsaem.9b00038

[20] Soto Pérez JJ, Peña-Duarte A, Trinidad P, Quintana G, Zhao X, Sasaki K, et al. Combined rotating disk slurry electrodeposition–spontaneous galvanic displacement for Pt-M (M = Co, Ni, and Cu) catalyst synthesis for the oxygen reduction reaction in alkaline media. ACS Applied Energy Materials. 2022;**5**(12):15175-15187. DOI: 10.1021/acsaem.2c02843

[21] Santiago D, Rodríguez-Calero GG, Rivera H, Tryk DA, Scibioh MA, Cabrera CR. Platinum electrodeposition at high surface area carbon Vulcan-XC-72R material using a rotating disk-slurry electrode technique. Journal of the Electrochemical Society. 2010;**157**(12):F189. DOI: 10.1149/1.3489948

[22] Mahmoud AS et al. Platinum and CeO_{2-x} Occlusion Electrodeposition at Unsupported Vulcan XC-72R for Methanol and Ethanol Oxidation Reactions in Alkaline Media. IOPscience. ECS Advances. 2024;**3**:044501. DOI: 10.1149/2754-2734/ad7bcb

[23] Tian, editor. High durability of Fe–N–C single-atom catalysts with carbon vacancies toward the oxygen reduction reaction in alkaline media. In: Advanced Materials. Wiley Online Library; 2023. DOI: 10.1002/adma.202210714

[24] Omoogun A, Chatterjee S, Uddin Ahmed S, Cabrera RC. Sequential Electrodeposition of Platinum-Nickel

on Boron Doped Diamond Electrode for Ammonia Oxidation Reaction. IntechOpen; 2025. DOI: 10.5772/intechopen.1008559

[25] Bhuvanendran N, Ravichandran S, Xu Q, Maiyalagan T, Su H. A quick guide to the assessment of key electrochemical performance indicators for the oxygen reduction reaction: A comprehensive review. International Journal of Hydrogen Energy. 2022;**47**(11):7113-7138. DOI: 10.1016/j.ijhydene.2021.12.072

[26] Stamenkovic VR, Fowler B, Mun BS, Wang G, Ross PN, Lucas CA, et al. Improved oxygen reduction activity on Pt_3 Ni(111) via increased surface site availability. Science. 2007;**315**(5811):493-497. DOI: 10.1126/science.1135941

[27] Jesus EC, Santiago D, Casillas G, Mayoral A, Magen C, José-Yacaman M, et al. Platinum electrodeposition on unsupported single wall carbon nanotubes and its application as methane sensing material. Journal of the Electrochemical Society. 2012;**160**(2):H98. DOI: 10.1149/2.054302jes

[28] Li, editor. Fe–Ni alloy nanoclusters anchored on carbon aerogels as high-efficiency oxygen electrocatalysts in rechargeable Zn–air batteries. In: Small. Wiley Online Library; 2021. DOI: 10.1002/smll.202102002

[29] Chatterjee S, Cabrera CR. Rotating disk slurry electrodeposition of trimetallic nickel-cobalt-iron nanoparticles electrocatalyst on Vulcan xc-72R for the oxygen reduction reaction in alkaline media. ECS Meeting Abstracts. 2024;**MA2024-01**(53):2855. DOI: 10.1149/MA2024-01532855mtgabs

[30] Namvari M, Chakrabarti BK. Electrophoretic deposition of MXenes and their composites: Toward a scalable approach. Advances in Colloid and Interface Science. 2024;**331**:103208. DOI: 10.1016/j.cis.2024.103208

[31] Yang Y, Peng H, Xiong Y, Li Q, Lu J, Xiao L, et al. High-loading composition-tolerant Co–Mn spinel oxides with performance beyond 1 W/Cm^2 in alkaline polymer electrolyte fuel cells. ACS Energy Letters. 2019;**4**(6):1251-1257. DOI: 10.1021/acsenergylett.9b00597

[32] Vlamidis Y, Scavetta E, Gazzano M, Tonelli D. Iron vs aluminum based layered double hydroxides as water splitting catalysts. Electrochimica Acta. 2016;**188**:653-660. DOI: 10.1016/j.electacta.2015.12.059

[33] Shi Y, Huang W-M, Li J, Zhou Y, Li Z-Q, Yin Y-C, et al. Site-specific electrodeposition enables self-terminating growth of atomically dispersed metal catalysts. Nature Communications. 2020;**11**:4558. DOI: 10.1038/s41467-020-18430-8

[34] Cui X, O'Hayre R, Pylypenko S, Zhang L, Zeng L, Zhang X, et al. Fabrication of a mesoporous $Ba_{0.5}Sr_{0.5}Co_{0.8}Fe_{0.2}O_{3-\delta}$ perovskite as a low-cost and efficient catalyst for oxygen reduction. Dalton Transactions. 2017;**46**(40):13903-13911. DOI: 10.1039/C7DT03082G

[35] Baz FB, El Sehiemy RA, Bayoumi ASA, Abaza A. Parameter extraction of proton exchange membrane fuel cell based on artificial rabbits' optimization algorithm and conducting laboratory tests. Scientific Reports. 2024;**14**(1):21145. DOI: 10.1038/s41598-024-70886-6

[36] Alias MS, Kamarudin SK, Zainoodin AM, Masdar MS. Active direct methanol fuel cell: An overview. International Journal of Hydrogen

Energy. 2020;**45**(38):19620-19641. DOI: 10.1016/j.ijhydene.2020.04.202

[37] Zhu J, Pedersen A, Kellner S, Hunter RD, Barrio J. Impact of ionomers on porous Fe-N-C catalysts for alkaline oxygen reduction in gas diffusion electrodes. Communications Chemistry. 2025;**8**(1):27. DOI: 10.1038/s42004-025-01422-4

[38] Facet impact of $CuMn_2O_4$ spinel nanocatalysts on enhancement of the oxygen reduction reaction in alkaline media. ACS Catalysis. DOI: 10.1021/acscatal.2c03275

[39] Liu K, Fu J, Lin Y, Luo T, Ni G, Li H, et al. Insights into the activity of single-atom Fe-N-C catalysts for oxygen reduction reaction. Nature Communications. 2022;**13**(1):2075. DOI: 10.1038/s41467-022-29797-1

Chapter 4

Synthesis of Catalysts Using RF Magnetron Sputtering for the Electrochemical Reduction of Carbon Dioxide Gas to Ethylene Gas

Magdalena Wajrak and Kurt L. Macri

Abstract

Carbon dioxide (CO_2) gas is a main contributor to green-house effect and is the driver of the climate change. The catalytic electrochemical reduction of CO_2 (eCO_2RR) to produce value-added molecules is now emerging as a viable strategy to control the increasing concentration of atmospheric CO_2. For this process, the selection of an appropriate catalyst is crucial in determining product selectivity. Hence, there is substantial research dedicated to identifying catalysts that preferentially produce specific value-added molecules. Copper has been identified as having strong selectivity toward ethylene production. This chapter outlines the viability of the radio frequency (RF) sputtering technique as a novel approach to the development of copper catalysts to selectively produce ethylene gas using H-type electrochemical cell. Key variables, including substrate type, film thickness, substrate temperature, and co-sputtering with other metals were investigated to optimize the efficiency of copper sputtered catalysts for ethylene synthesis. The results demonstrate that the RF sputtering method for catalyst fabrication is effective in promoting ethylene production, achieving a Faradaic efficiency of almost 20% (19.5%) at −2400 mV vs. Ag/AgCl with a 60 nm copper sputtered film at RF = 150 W on a gas diffusion electrode paper substrate held at room temperature.

Keywords: greenhouse gas, climate change, electrochemistry, CO_2 reduction, catalysts, RF sputtering

1. Introduction

Climate shift is one of the most critical and serious threats that the world is currently facing, and most in need of action. Climate change is caused primarily by the production and emission of greenhouse gasses, which is a term used to describe a variety of naturally occurring and synthetic gasses such as sulfur and nitrogen oxides, ozone, carbon dioxide and halocarbons [1]. While excess greenhouse gasses are responsible for the warming climate, complete removal of these gasses is not desirable as appropriate levels of these greenhouse gasses regulate the global temperature allowing for life. Problems arise when the concentrations of these gases increase as this leads to an increase in the greenhouse effect, trapping more of this terrestrial radiation leading to increasing temperatures. The concentration of CO_2 gas has increased from approx. 315 ppm (March 1958) to 427 ppm (February 2025), measured by the Mauna Loa Observatory [2, 3], which gives an average increase of 1.67 ppm per year. However, in 2024 there was an almost 3 ppm increase from 2023, which was the largest annual increase (**Figure 1**).

This is alarming because the increasing levels of greenhouse gasses are directly related to the increasing global temperature and rising greenhouse gas emissions will continue to cause a further increase in global temperatures.

The main drivers of the increase of CO_2 concentration come from the power sector with an increase of 261 megatons, and transport sector with an increase of 254 megatons over the course of 1 year (2021–2022). These values offset the decreasing levels of emissions in the industry and building sectors, thus resulting in an overall steady emission increase [5]. Consequently, several global government policies have been implemented to mitigate the rising greenhouse gas emissions. The most notable being the Kyoto Protocol and the Paris Agreement [6]. While there are policies in place, reducing CO_2 emissions is proving to be challenging in the current industrial world.

1.1 Novel approach to reducing amount of CO_2 in the atmosphere

A possible solution to the CO_2 problem is to convert the CO_2 gas it into "value-added" molecules, such as ethylene or methanol, via catalytic electrochemical reduction (**Figure 2**).

The electrochemical CO_2 reduction reaction (eCO_2RR) requires considerable amount of energy and a catalyst due to fact that each oxygen atom is doubly bonded to the carbon atom, making the CO_2 molecule relatively stable. The purpose of the catalyst is to provide a site for CO_2 binding and promoting the C=O bond breaking and facilitate reactions to form other molecules [8]. The type of catalyst is instrumental in

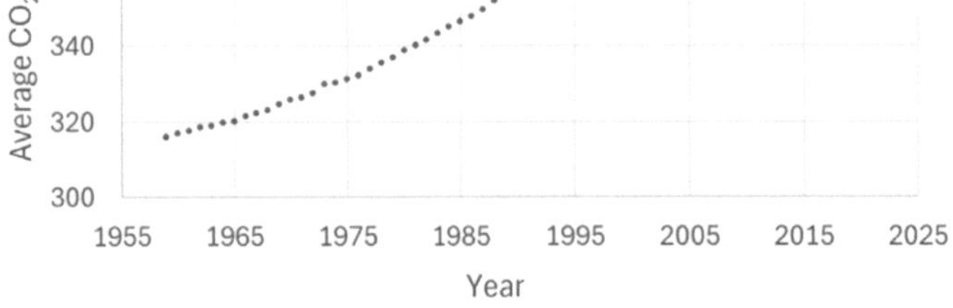

Figure 1.
The full record of average yearly CO_2 concentrations since 1958 (graph generated by the authors from data provided by Lan and Keeling [4]).

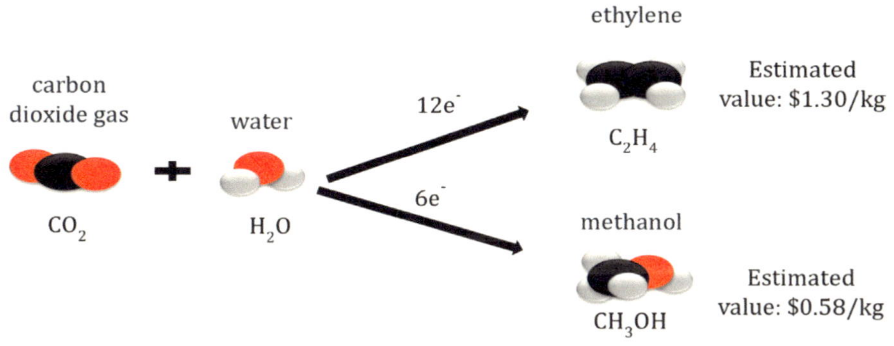

Figure 2.
Two possible reaction pathways for CO_2 reduction to form methanol or ethylene and the unit price of each of those molecules (image generated by the authors and data provided by Song et al. [7]).

the formation of products, for example a copper-based catalyst preferentially facilitates the formation of ethylene molecule, whereas a tin catalyst favors the production of formic acid [9]. The ability to selectively direct the production and efficiency of specific products by adjusting the catalyst and electrochemical parameters enhances the appeal and importance of this process.

1.2 Designing innovative catalysts

In the electrochemical reduction of CO_2, it has been shown that the catalyst is the most important factor in driving the efficiency, selectivity, and energy requirements for the CO_2 conversion process [8]. Consequently, most of the research effort has gone into the development of catalysts and understanding of the mechanism of conversion of CO_2 to other molecules [10]. However, what has come to light from literature review, is that a well-known technique for creating thin films for semi-conductor industry, nanotechnology and optical science, has not been fully investigated and utilized for the development of catalysts for eCO_2RR. Sputtering is ideal for the production of thin films and manipulating surfaces and thus could potentially offer a valuable alternative to current catalyst designs.

This chapter presents research which employed a particular type of sputtering technique, Radio Frequency (RF) Magnetron Sputtering, to fabricate copper-based catalysts and assess their potential to form the highest (in terms of price per unit) value-added molecule, ethylene. The sputtering method offers enhanced flexibility in catalyst development by allowing precise control over film thickness, deposited particle size, and the incorporation of multiple sputtered elements and other gaseous dopants during the deposition process. Additionally, sputtering permits the use of various substrates, further expanding the range of possible catalyst configurations. Hence, this work presents a preliminary fundamental study of the use of the sputtering technique in the design of catalysts to produce ethylene gas and optimization of those catalysts to improve ethylene selectivity by measuring Faradaic efficiency (FE) in H-type electrochemical cell.

1.3 Copper catalyst for ethylene production

Copper has been shown to be the most selective catalyst for the synthesis of ethylene from CO_2 [11]. Some of the earliest work using a copper catalyst for the formation of ethylene in a weakly acidic solution was done by Hori [11]. Hori found that the reduction process has reasonably high yields of between 30% and 48%, however, the selectivity for ethylene alone was rather poor. Another work [12] reported FE of 22% for ethylene when using polished metallic copper and a maximum FE of 60% when using a copper foil treated with an O_2 plasma. However, further increasing the O_2 treatment resulted in a decline in selectivity, possibly due to changing of the copper surface. This work once again demonstrates that a copper-based catalyst does have a selective preference for producing ethylene gas, unfortunately, there was no long-term testing of the catalyst. Exploring the stability of catalysts is crucial, since if the catalyst is highly selective for only a short period of time, it would not make it viable for industrial applications. In 2017, Jeon et al. [13] also showed that a copper-based catalyst results in the formation of ethylene and in this work, stability experiments were performed. The copper catalyst was stable for over 12 hours of operation and the efficiency of the catalyst did not decrease over this time.

What is intriguing is that none of the above-mentioned research implemented the use of a sputtering technique for the design of a catalyst in the production of ethylene and considering that despite achieving relatively high efficiencies for ethylene production, it is still challenging to selectively generate one specific product and to obtain long term stability without the use of complicated catalysts.

1.4 Importance of catalyst synthesis

The review article by Qu et al. [8] demonstrates that the design of copper-based electrocatalysts for ethylene production with high energy density and selectivity is crucial for the advancement of CO_2 reduction reaction to ethylene. The authors propose that "the preparation of catalysts and surface engineering at the atomic level can profoundly influence the interface microenvironment, thereby impacting the selectivity of CO_2 conversion to ethylene." This is because, it has been shown that "the binding energies of reaction intermediates" are highly dependent on the surface type and structure.

The work of Das et al. [14] suggested that a variation in the temperature of the substrate, resulted in a variation in the crystallite structure. It showed that with higher temperatures larger crystal formation was observed on the substrate. At lower temperatures, there was better adhesion and a more uniform coating and thickness. It was also found that a higher temperature results in a slightly lower conductivity, a large decrease on resistivity, and an increase in electron mobility. Another article [15] explained that increasing the length of time that the sputtering takes place, results in a thicker layer of deposition and crystallite size. This in turn will lead to a larger surface area for the catalyst. Work by Tabassum et al. [16] suggested that varying the size of the sputtered particles results in a higher surface area for the reactions to occur and this in turn could lead to improved catalytic activity which would result in higher efficiency. Finally, Lai et al. [17] suggested that a variation of the substrate will result in the deposition of the material being varied, such as the size of the grains or how closely they packed together, which can further increase the efficiency of the catalyst.

Consequently, given the limited research on the use of sputtering technique to design catalysts specifically for ethylene production, there is opportunity to assess the viability of this method and investigate sputtering parameters which are most conducive to the efficient production of ethylene gas.

2. Sputtering theory

The sputtering process occurs inside a vacuum chamber (**Figure 3**—left), which is pumped down to high vacuum to remove any "foreign" atoms/molecules that could interfere with the sputtering process. The chamber contains a target, which is the material that is used to deposit onto the substrate. The chamber is supplied with an inert gas such as argon and when an RF field is generated at the target, the argon gas is ionized forming plasma. Once plasma has been generated in the system, the pressure and flow rate of the inert gas is reduced to increase the mean free path of the deposition material. The mean free path is the average distance the deposition material travels before impacting another ion or particle in the chamber. The ionized gas is attracted to the negatively charged target material (**Figure 3**—center) and the incident ions impact the surface with sufficient energy to dislodge atoms from the surface (**Figure 3**—right) and deposit these atoms onto the substrate resulting in a formation of a thin film [19].

2.1 Radio frequency magnetron sputtering

In this research Radio Frequency Magnetron sputtering system was used to generate copper catalysts. Magnetron sputtering method requires application of a magnetic field around the target material. The magnetic field serves to prevent the electrons generated by the collision of positive ions with the negatively charged target surface from colliding with the substrate. This minimizes the loss of deposition material and thus enhances deposition rate [20]. The dislodged atoms are subsequently deposited onto the substrate, forming a uniform coating of the target element [21]. **Figure 4** depicts the RF magnetron sputtering process, where the substrate is positioned above the sputtering target, which is not specific to RF Magnetron sputtering systems. The position of substrate and target can vary with each sputtering system.

RF magnetron sputtering can also be combined with a reactive sputtering method, which involves introducing a reactive gas, such as oxygen or nitrogen, into the chamber along with the inert argon. When the argon gas is ionized, the reactive gas is also ionized

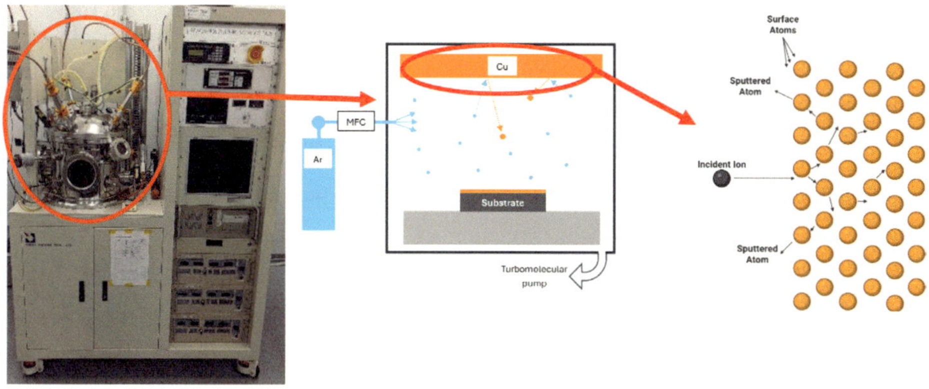

Figure 3.
(Left) Korea Vacuum Tech R.F. magnetron sputtering system used at ECU (photo taken by the authors), (center) schematic illustration of the essential features of sputtering (image generated by the authors, adopted from Greene [18]) and (right) removal of target material during the sputtering process (image generated by the authors using BioRender, adopted from Rossnagel [19]).

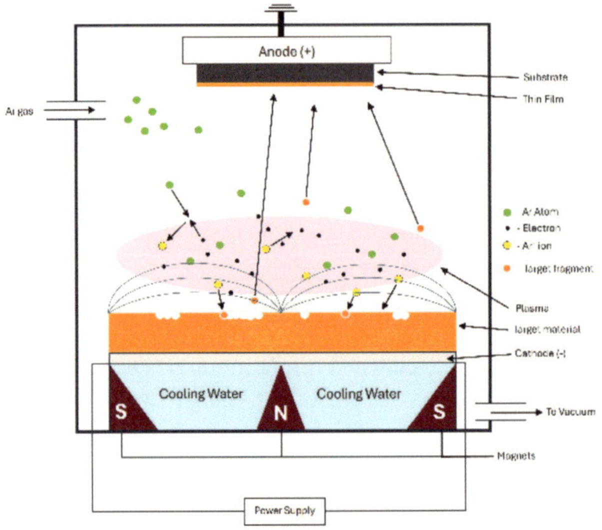

Figure 4.
Schematic of RF magnetron sputtering device (image generated by the authors, adopted from Hishimone et al. [22]).

and upon impact with the target material, the ions from both gases react, leading to the formation of oxide or nitride on the substrate thus creating tailored films. It is also possible to sputter with two or more targets creating even more unique films.

2.2 Effect of sputtering parameters on film characteristics

Film characteristics, such as thickness, size of particles, and roughness are closely linked to the sputtering parameters such as RF field strength and inert gas pressure. Studies have shown [23, 24] that the rate of deposition and therefore the deposition thickness is affected by the following sputtering parameters; the pressure of the chamber, the pressure of the inert gas in the system, the density of the target material, and the RF power supplied to the target. Decreasing the pressure in the chamber (i.e., improved vacuum) means that there are less particles in the chamber which increases the mean path length of the dislodged deposition material. The pressure of the inert gas also affects the deposition rate [24]. An increased gas load in the system leads to increased ionization and thus more particles colliding with the target material, however, the higher the concentration of the inert gas, the lower the mean free path for the deposited material.

The density of the target material also plays a significant role in determining the deposition rate [25]. A denser target requires less energetic ions to dislodge particles from its surface during the sputtering process. Additionally, a higher density means that more target material particles are ejected per ion impact, which consequently increases the deposition rate. Varying the RF power is utilized to increase the deposition rate; as the RF power increases, the ions become more energetic and impact the target with greater frequency and energy, leading to a higher deposition rate.

Monitoring deposition rate is therefore particularly important when sputtering thin films, because it determines film thickness and allows for production of reproducible films. Throughout the sputtering process, sputtering software is employed to

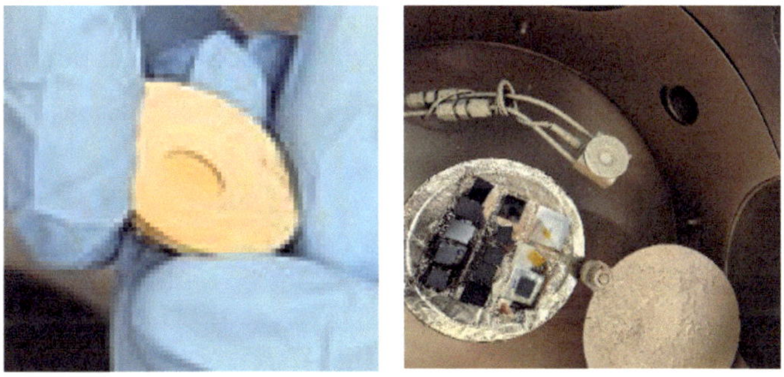

Figure 5.
Gold coated quartz crystal (left) and position of the crystal in the vacuum chamber in the RF magnetron sputtering system used in this project (right) (photos taken by the authors).

monitor and calculate the deposition rate in real-time. A gold-coated quartz crystal (**Figure 5**), which is deposited simultaneously with the substrate serves as a reference for determining the deposition rate.

3. Electrochemical reduction of CO_2 gas

3.1 H-type cell

Electrochemical reduction of CO_2 gas is conducted in a H-type cell (**Figure 6**), featuring two chambers separated by glass frit or a semi-permeable membrane and three electrodes [27].

These chambers are filled with an electrolyte solution and become the sites for the individual half-cell reactions. The purpose of these types of cells is to ensure that the working electrode is separated from the counter electrode, as the membrane between the cells only allows for ion transfer. The membranes can be changed depending on the specific type of ion species. H-type cells are typically used in cases where the potential needs to be measured with a high level of accuracy. It is also predominantly used for generation of gasses through the reduction process. Although there is energy

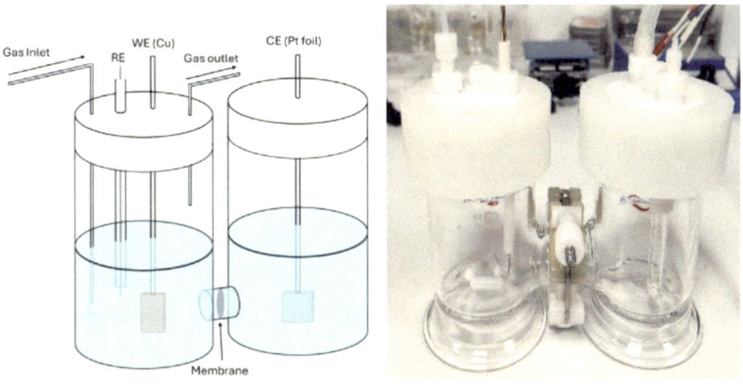

Figure 6.
(Left) Sketch of a H-type cell (image generated by the authors, adopted from Alinejad et al. [26]), (right) photo taken by the authors of the actual H-type cell used in this project.

supplied in this process, the CO_2 reduction in a H-type cell is classified as an electrochemical process, because the application of voltage in this case is purely to aid the reduction of CO_2, due to stability of CO_2.

3.2 Electrochemistry of ethylene formation

Ethylene production involves reduction of CO_2 at the cathode and at the same time oxidation of H_2O at anode (**Figure 7**), although the actual formation of ethylene involves several pathways, the overall reaction is:

$$\text{Cathode}: 2CO_{2(aq)} + 8H_2O_{(l)} + 12e^- \rightarrow C_2H_{4(g)} + 12OH^-_{(aq)} \quad (1)$$

$$\text{Cathode (competing reaction)}: 2H_2O_{(l)} + 2e^- \rightarrow H_{2(g)} + 2OH^- \quad (2)$$

$$\text{Anode}: 4OH^-_{(aq)} \rightarrow O_{2(g)} + 2H_2O_{(l)} + 4e^- \quad (3)$$

$$\text{Overall}: 2CO_{2(g)} + 2H_2O_{(l)} \rightarrow C_2H_{4(g)} + 3O_{2(g)} \quad (4)$$

In the H-type cell the CO_2 gas is introduced to the cathode electrode, which is composed of a copper-based catalyst, where it is reduced to ethylene (C_2H_4) as the electrode is maintained at a negative potential. At the anode, water is oxidized to produce oxygen gas. However, due to the high stability of CO_2, the reduction process necessitates a high overpotential to activate CO_2 into the CO_2^{-*} anion radical, which occurs at −1.9 V relative to SHE. The situation is further complicated by the formation of hydrogen gas (H_2) at the cathode, as the reduction potentials of H_2 and CO_2

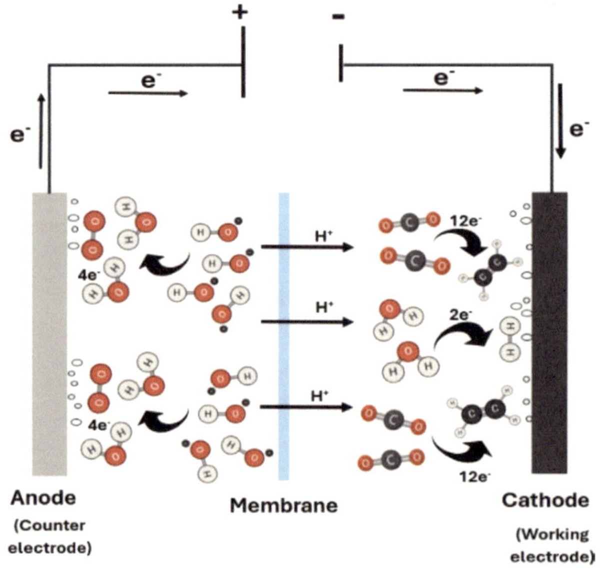

Figure 7.
Schematic of the reduction of CO_2 inside a H-type cell illustrating anode and cathode sites and movement of ions and electrons (image generated by the authors, adopted from Zhang et al. [28]).

are similar. This results in a competitive reaction between the two, with energy being diverted toward H_2 production instead of ethylene, thereby reducing the overall efficiency of the process. Consequently, the role of the catalyst becomes critical, as it must be selective enough to favor ethylene production while minimizing H_2 formation.

The suitability of a catalyst for ethylene production is determined by calculating the FE, which is equal to the overall charge that is needed to produce the specific product of interest, in this case, ethylene, and dividing by the total charge produced during the electrochemical reduction reaction and multiplying by 100 for percentage [29];

$$FE = \left[\frac{Q_{product}}{Q_{total}}\right] \times 100 = \frac{Z \times F \times \phi \times c}{V_m \times i} \qquad (5)$$

where $Z \times F \times \Phi / Vm$ are all constant for each experimental run. Thus, the only parameters that will vary with each experiment are i and c. Current can be obtained from running chronoamperometry (CA) run at the start of each reduction reaction and concentration will be obtained from the area under the chromatographic peak of ethylene.

3.3 Chronoamperometry and cyclic voltammetry

Calculating the FE of CO_2 reduction also requires two additional pieces of electrochemical information; determination of the potential for the reduction process and monitoring of the current over the duration of the experiment whilst constant potential is applied [29]. Cyclic voltammetry (CV) is completed over a range of potentials to establish the relationship between the current and voltage and decide on the appropriate potential for the reduction process. CA is performed by applying a constant potential at a voltage determined from the CV run to the working electrode and measuring the resulting current over time. Monitoring the current response over time when a constant potential is applied allows the evaluation of the efficiency of the catalyst to facilitate CO_2 reduction and whether the current stabilizes or decreases over time, providing information regarding catalyst's stability and effectiveness. From the CA analysis, current values can be obtained and integrated over the duration of the reduction reaction or taken at a particular time of reaction, if current remains stable, providing Q_{total} value needed for FE calculation (Eq. (5)).

4. Methodology

The initial phase involved the synthesis of the catalysts using sputtering technique, which comprised of preparing the RF magnetron sputtering system, loading the required sputtering target or targets, placing the substrates on stage, creating a vacuum in the chamber and sputtering for specific time to reach desired film thickness as shown in **Figure 8**.

Phase 2 involved testing the catalysts developed in phase 1 using H-type electrochemical cell [9–11] and analyzing the gaseous products of the electrochemical reduction using Gas Chromatography with Flame Ionization and Thermal Conductivity Detectors (FID/TCD) instrument. A visual representation of phase 2 is presented in **Figure 9**.

Figure 8.
Steps involved in phase 1 (from left to right): preparing the sputtering system, installation of target, mounting substrates inside the vacuum chamber and sputtered substrates with specific film thickness (photos taken by the authors).

Figure 9.
Steps involved in phase 2 (from left to right): mounting the synthesized catalyst in the working electrode holder, testing catalysts using H-type electrochemical cell, analysis of gaseous products with GC instrument (photos taken by the authors and GC chromatograph generated by the GC software).

The final phase involved characterization of the surface of the catalysts prior to sputtering, after sputtering and after electrochemical testing using Hitachi SU3500 SEM and Tescan Clara SEM. EDX and XPS analyses were also performed on some samples. Overview of the three phases of the project is shown in **Figure 10**.

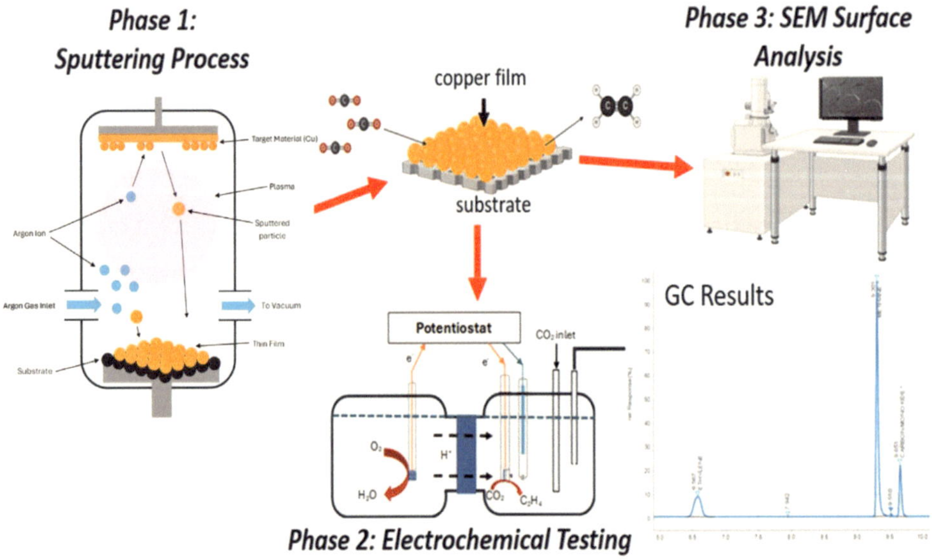

Figure 10.
Illustration of the overview of the three phases (from left to right): sputtering various substrates, testing catalysts using H-type electrochemical cell, analysis of gaseous products with GC instrument and surface characterization using SEM (illustration and images generated by the authors using BioRender, Agilent GC software and sputtering image adopted from [30]).

5. Electrochemical testing of catalysts

5.1 Substrate type investigation

The effect of the substrate on ethylene production was investigated with five different substrates (glass substrate was not tested). Unsputtered substrates were tested first to determine the contribution of the substrate itself to the formation of ethylene. The results are presented in **Table 1**. As expected, pure copper foil produced the largest amount of ethylene with FE = 10.1%, interestingly the GDE paper also formed small amount of ethylene with FE = 0.7%, however, the rest of the substrates as expected did not. The GDE paper consists of two sides, one is made from carbon rod-shaped structure to provide electrical conductivity. The other side is made of carbon-based material with high porosity, providing a large surface area for the electrochemical reactions to occur and encourages mass transport of gases within the electrochemical cell. This side is also impregnated with polytetrafluoroethylene (PTFE) to increase hydrophobicity to prevent water from flooding the electrode, which can hinder gas diffusion and reduce efficiency. Although carbon itself is not particularly efficient at catalyzing CO_2, carbon nanotubes have been used as catalysts for CO_2 [8], Hence, GDE could produce small amount of ethylene.

All six substrates were sputtered with 60 nm copper and electrochemically tested (**Figure 11**). Results are presented in **Table 2**.

Substrate type	Current (A)	Faradaic efficiency (%)
Titanium foil	0.035	0.0
Copper foil	0.035	10.1
Silicon wafer	0.006	0.0
GDE paper	0.029	0.7
Carbon paper	0.028	0.0

Table 1.
Electrochemical testing of unsputtered substrates from data obtained from GC by Wajrak at −2400 mV.

Figure 11.
Photo (taken by the authors) of CO_2 gas being bubbled on the left side directly onto copper sputtered GDE substrate being held in place by working electrode holder in a H-cell and being reduced to ethylene (60 nm, −2400 mV, RF = 150 W).

Substrate	Current (A)	C_2H_4 concentration from peak area compared to gas standard (ppm)	FE (%)
Glass	0.0425	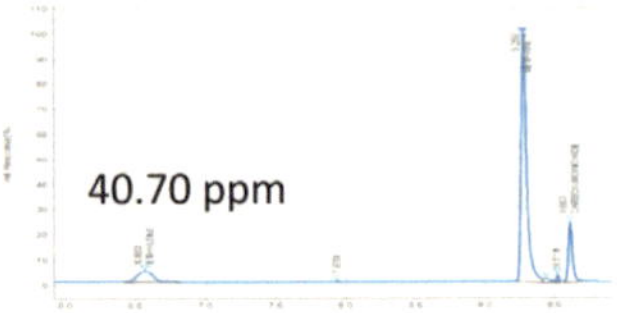	2.5
Carbon paper	0.0276	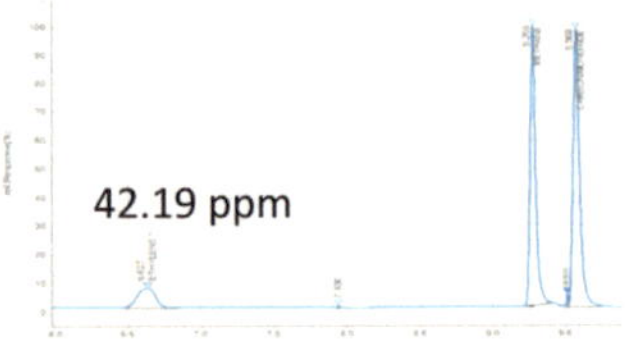	4.1
Titanium foil	0.031	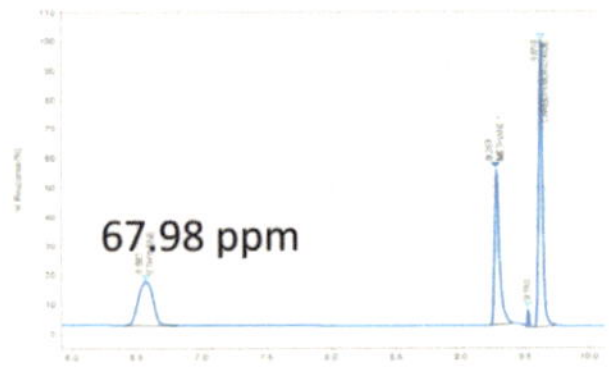	5.8
Silicon wafer	0.0343	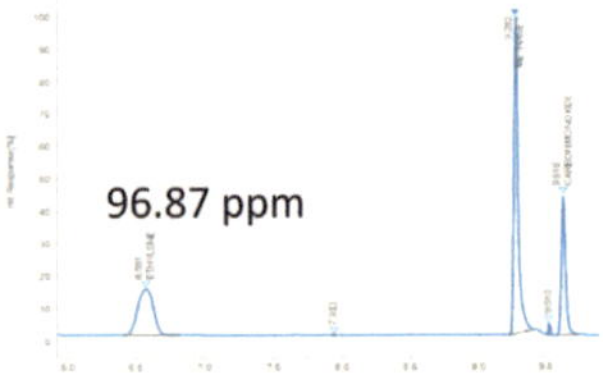	7.4
Copper foil	0.041	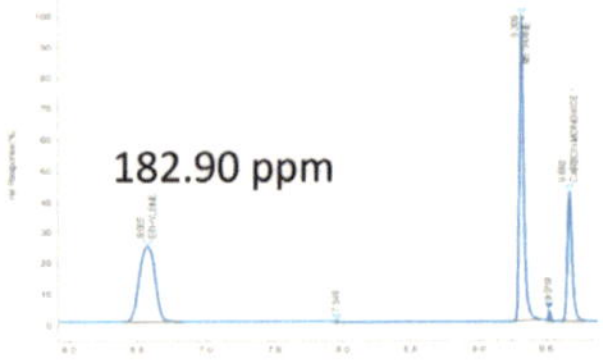	11.7
GDE	0.0386	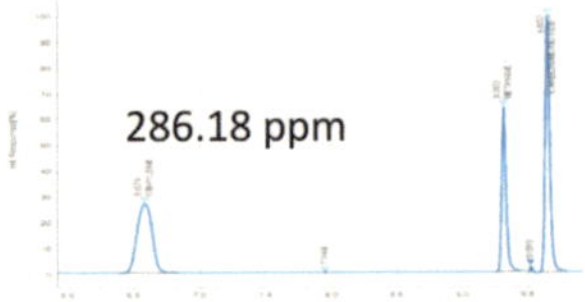	19.5

Table 2.
Copper sputtered on various substrates from data obtained from GC by the authors (RF = 150 W, 60 nm, −2400 mV).

While all copper sputtered catalysts, even glass substrate generated ethylene, GDE outperformed all the other substrates in ethylene production. This was an expected result since the GDE paper, as mentioned before, is specifically designed for gas electrochemical reactions by enhancing gas diffusion and the PTFE coating increases hydrophobicity eliminating water thus maintaining porosity of the surface and allowing more CO_2 interactions with catalyst. The copper foil was the next best substrate after GDE, which again was expected since there is already higher concentration of copper making up the catalyst and therefore more opportunity for CO_2 molecules to interact with the substrate and thus convert to ethylene.

5.2 Target material investigation

Two target materials were investigated; copper (99.9999%) and copper oxide (99.95%). Results are presented in **Table 3**. It was interesting to see that copper oxide FE was almost half of the value of pure copper. Although, more comprehensive investigation is required here, these results clearly demonstrate the significant effect of purity of copper sputtered film on ethylene production.

5.3 Film thickness investigation

Literature [31] has shown that the thickness of the catalyst surface influences the efficiency of ethylene production. To investigate this effect, copper films with thicknesses of approximately 10, 60, and 400 nm were examined (**Table 4**). The findings indicated that thinner films resulted in nearly a 3% improvement in ethylene efficiency. This preference for thinner films may be attributed to the fact that thicker films could obscure the catalyst's surface topography by filling the pores within the gas diffusion electrode (GDE) paper, thereby reducing the reactive surface area. Additionally, thicker films exhibited retention issues, which further compromised their catalytic performance. The results demonstrate that thinner films generally perform better, however, more film thickness values should be analyzed.

5.4 Radio frequency power investigation

Three RF power values were investigated with results shown in **Table 5**. There is an almost 5% increase in ethylene FE going from 75 to 300 W RF power. This is almost

Targets	Current (A)	C_2H_4 peak area compared to standard (ppm)	FE (%)
Cu (99.9999%)	0.041	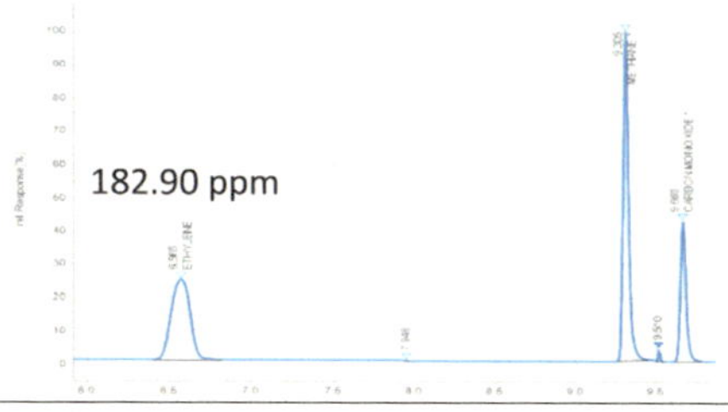	11.7
CuO (99.95%)	0.043	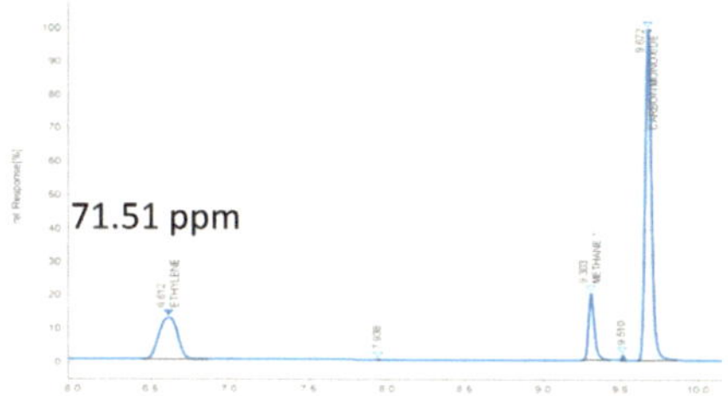	6.7

Table 3.
Various targets sputtered on Cu foil from data obtained by the authors from GC (RF = 150 W, 60 nm, −2400 mV).

Film thickness (nm)	Current (A)	C_2H_4 peak area compared to standard (ppm)	FE (%)
400 nm	0.050	287.218	15.0
60 nm	0.035	237.916	17.9
10 nm	0.028	164.807	15.7

Table 4.
Various Cu film thickness sputtered on GDE paper from data obtained from GC and H-cell (RF = 150 W, −2400 mV).

RF power (W)	Current (A)	C_2H_4 peak area compared to standard (ppm)	FE (%)
75	0.028	144.285	13.6
150	0.035	237.916	17.9
300	0.029	207.465	18.8

Table 5.
Variation of the RF power for Cu film sputtered on GDE from data obtained from GC and H-cell (60 nm, −2400 mV).

certainly due to the change in morphology of the sputtered film, where the increase in RF power produced significantly more and larger copper nanograin clusters. This agrees with what would be expected [32] since the size of sputtered grains depends on deposition temperature, sputtering power, pressure, and the material being sputtered. In most cases, the sputtered material forms polycrystalline films with nanometer-sized grains. Larger nano-grain clusters result in a greater surface roughness [33] which in turn will provide larger surface area and more binding for CO_2 molecules and thus more opportunity for reaction [34]. Although using higher RF power to generate copper film does increase production of ethylene, the increased spiking in the deposition rate is not desirable, due to uneven film formation. Therefore, an RF value of 150 W was chosen, because it gave stable deposition rate and only less than 1% variation in FE from the 300 W RF power.

5.5 Optimized parameters

The investigation of various sputtering and electrochemical parameters has clearly demonstrated that the RF magnetron sputtering is a viable method for producing catalysts for CO_2 reduction to ethylene gas. These preliminary results indicate that maximum ethylene production efficiency of almost 20% is obtained with a high purity (99.9999%) copper target sputtered on GDE on both sides to a thickness of approximately 60 nm using RF power of 150 W.

6. Catalyst development overview and discussion

The most effective target material for ethylene formation is high purity copper. This is attributed to copper's ability to provide the optimal binding sites for the CO_2 reduction process. The thickness of the catalyst film also had a noticeable impact on ethylene selectivity. Thicker films were generally associated with lower FE values.

Specifically, GDE carbon paper with a 60 nm Cu film outperformed by almost 3% both the much thicker film (400 nm) and thinner film (10 nm). This decrease in efficiency is most likely due to the reduction in surface morphology as the thicker coating "fills-in" the substrate's existing morphology, leading to a decrease in available surface area for catalytic reactions. Finally, the RF power had the most significantly effect on the FE, going from 75 to 300 W increases the efficiency by more than 5%. The higher RF power resulted in substantial increase in surface morphology, thereby offering more surface area for catalytic interactions.

7. Future work

The application of RF magnetron sputtering to produce films for the electrochemical reduction of CO_2 to ethylene has emerged as a promising technique for advancing research in CO_2 reduction. Although several parameters have been investigated here, there are many more factors that need to be further examined, this project has only "scratched the surface" of this research field. Future work will include; analysis of catalysts with copper film thickness of less than 60 nm, optimizing the RF power and annealing methods for improved film adhesion, investigating various electrolyte types and co-sputtering with other elements and compounds to increase faradaic efficiency.

The long-term goal of this research is to develop a method capable of efficiently reducing CO_2 into valuable molecules on a larger scale.

Ultimately, the objective is to implement these sputtered catalysts in industrial-scale technologies to help reduce CO_2 emissions and mitigate global warming (**Figure 12**).

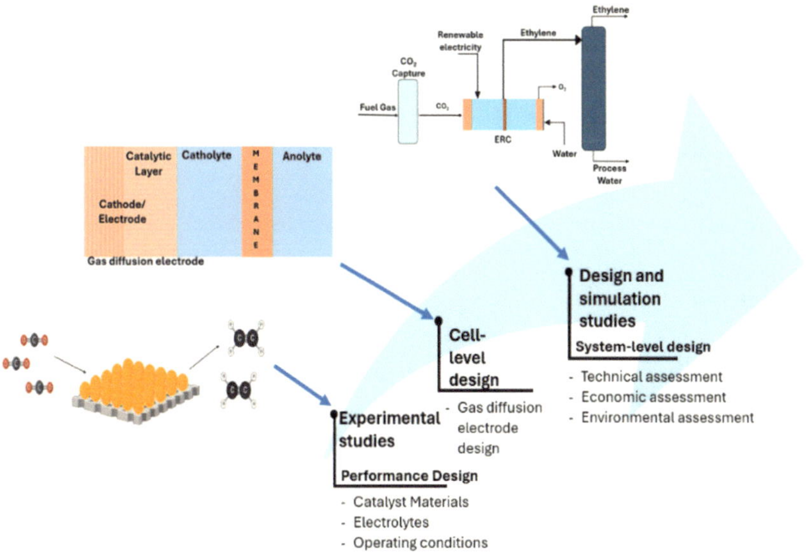

Figure 12.
A schematic illustrating how this project integrates into the future research of the electrochemical CO_2 reduction reaction (image generated by the authors, adopted from Wiranarongkorn [35]).

Conflict of interest

The authors declare no conflict of interest.

Author details

Magdalena Wajrak* and Kurt L. Macri
Edith Cowan University, Perth, Australia

*Address all correspondence to: m.wajrak@ecu.edu.au

References

[1] Ritchie H, Roser M, Rosado P. CO_2 and Other Greenhouse Gas Emissions. Our World in Data; 2020 [Internet]. Available from: https://ourworldindata.org/co2-and-other-greenhouse-gas-emissions [Accessed: May 13, 2025]

[2] US Department of Commerce N. Global Monitoring Laboratory—Carbon Cycle Greenhouse Gases [Internet]. 2020. Available from: https://gml.noaa.gov/ccgg/about/co2_measurements [Accessed: November 23, 2024]

[3] US Department of Commerce N. Global Monitoring Laboratory—Carbon Cycle Greenhouse Gases [Internet]. Available from: https://gml.noaa.gov/ccgg/trends/mlo.html [Accessed: November 23, 2024]

[4] Lan X, Keeling R. Trends in Atmospheric Carbon Dioxide (CO_2) [Internet]. 2024. Available from: https://gml.noaa.gov/ccgg/trends/data.html [Accessed: July 10, 2025]

[5] International Energy Agency. CO_2 Emissions in 2022. IEA; 2023 [Internet]. Available from: https://www.iea.org/reports/co2-emissions-in-2022 [Accessed: January 23, 2025]

[6] Paul L. Boley Law Library: International Climate Change Law Research Guide: International Agreements [Internet]. 2024. Available from: https://lawlib.lclark.edu/climate/treaties [Accessed: August 26, 2024]

[7] Song J, Song H, Kim B, Oh J. Towards higher rate electrochemical CO_2 conversion: From liquid-phase to gas-phase systems. Catalysts. 2019;**9**(3):224. DOI: 10.3390/catal9030224

[8] Qu J, Cao X, Gao L, Li J, Li L, Xie Y, et al. Electrochemical carbon dioxide reduction to ethylene: From mechanistic understanding to catalyst surface engineering. Nano-Micro Letters. 2023;**15**(1):178. DOI: 10.1007/s40820-023-01146-x

[9] Yu F, Deng K, Du M, Wang W, Liu F, Liang D. Electrochemical CO_2 reduction: From catalysts to reactive thermodynamics and kinetics. Carbon Capture Science & Technology. 2023;**6**(1):100081. DOI: 10.1016/j.ccst.2022.100081

[10] Chen X, Zhao Y, Han J, Bu Y. Copper-based catalysts for electrochemical reduction of carbon dioxide to ethylene. ChemPlusChem. 2023;**88**(1):e202200370. DOI: 10.1002/cplu.202200370

[11] Hori Y, Murata A, Takahashi R, Suzuki S. Enhanced formation of ethylene and alcohols at ambient temperature and pressure in electrochemical reduction of carbon dioxide at a copper electrode. Journal of the Chemical Society, Chemical Communications. 1988;**1**:17. DOI: 10.1039/c39880000017

[12] Mistry H, Varela AS, Bonifacio CS, Zegkinoglou I, Sinev I, Choi YW, et al. Highly selective plasma-activated copper catalysts for carbon dioxide reduction to ethylene. Nature Communications. 2016;**7**(1):12123. Available from: https://www. nature.com/articles/ncomms12123 [Accessed: May 6, 2024]

[13] Jeon HS, Kunze S, Scholten F, Roldan CB. Prism-shaped Cu nanocatalysts for electrochemical CO_2 reduction to ethylene. ACS Catalysis. 2017;**8**(1):531-535. DOI: 10.1021/acscatal.7b02959

[14] Das NK, Chakrabartty J, Farhad SFU, Sen Gupta AK, Ikball Ahamed EMK,

Rahman KS, et al. Effect of substrate temperature on the properties of RF sputtered CdS thin films for solar cell applications. Results in Physics. 2020;**17**:103132-103132. DOI: 10.1016/j.rinp.2020.103132

[15] Tchenka A, Agdad A, Samba Vall MC, Hnawi SK, Narjis A, Nkhaili L, et al. Effect of RF sputtering power and deposition time on optical and electrical properties of indium tin oxide thin film. Advances in Materials Science and Engineering. 2021;**2021**:1-14. DOI: 10.1155/2021/5556305

[16] Tabassum H, Yang X, Zou R, Wu G. Surface engineering of Cu catalysts for electrochemical reduction of CO_2 to value-added multi-carbon products. Chem Catalysis. 2022;**2**(1):1561-1593. DOI: 10.1016/j.checat.2022.04.012

[17] Lai CC, Boyd R, Svensson PO, Höglund C, Robinson L, Birch J, et al. Effect of substrate roughness and material selection on the microstructure of sputtering deposited boron carbide thin films. Surface and Coatings Technology. 2022;**433**:128160. DOI: 10.1016/j.surfcoat.2022.128160

[18] Greene JE. Review Article: Tracing the recorded history of thin-film sputter deposition: From the 1800s to 2017. Journal of Vacuum Science & Technology A: Vacuum, Surfaces, and Films. 2017;**35**(5):05C204. Available from: https://avs.scitation.org/doi/10.1116/1.4998940 [Accessed: June 14, 2024]

[19] Rossnagel S. Sputtering and Sputter Deposition. Handbook of Thin Film Deposition processes and Techniques. 2001;**2**(1):319-348. DOI: 10.1016/b978-081551442-8.50013-4

[20] Hughes M. What is Sputtering? Magnetron Sputtering? Semicore Equipment Inc.; 2014 [Internet]. Available from: https://www.semicore.com/what-is-sputtering [Accessed: June 14, 2024]

[21] Kelly PJ, Arnell RD. Magnetron sputtering: A review of recent developments and applications. Vacuum. 2000;**56**(3):159-172. DOI: 10.1016/s0042-207x(99)00189-x

[22] Hishimone PN, Nagai H, Sato M. Methods of Fabricating Thin Films for Energy Materials and Devices. London, UK: IntechOpen eBooks; 2020. DOI: 10.5772/intechopen.85912

[23] Aliramaji S, Keuter P, Neuß D, Hans M, Primetzhofer D, Depla D, et al. Effect of growth temperature and atmosphere exposure time on impurity incorporation in sputtered Mg, Al, and Ca thin films. Materials. 2023;**16**(1):414. DOI: 10.3390/ma16010414

[24] Rickerby DG, Serventi AM. Nanostructured metal oxide gas sensors for air-quality monitoring. Environonotechnology. 2010;**1**(1):99-136. DOI: 10.1016/b978-0-08-054820-3.00006-x

[25] Berg S, Katardjiev IV. Preferential sputtering effects in thin film processing. Journal of Vacuum Science & Technology A: Vacuum, Surfaces, and Films. 1999;**17**(4):1916-1925. DOI: 10.1116/1.581704

[26] Alinejad S, Quinson J, Li Y, Kong Y, Reichenberger S, Barcikowski S, et al. Optimizing the Use of a "Zero-Gap" gas diffusion electrode setup for electrochemical reduction of CO_2. ChemRxiv. 2021;**1**(1):8k376. DOI: 10.26434/chemrxiv-2021-8k376

[27] Pan H, Barile CJ. Electrochemical CO_2 reduction to methane with remarkably high faradaic efficiency

in the presence of a proton permeable membrane. Energy and Environmental Science. 2020;**13**(10):3567-3578. DOI: 10.1039/d0ee02189j

[28] Zhang W, Hu Y, Ma L, Zhu G, Wang Y, Xue X, et al. Progress and perspective of electrocatalytic CO_2 reduction for renewable carbonaceous fuels and chemicals. Advanced Science. 2017;**5**(1):1700275. DOI: 10.1002/advs.201700275

[29] Qiao Y, Kastlunger G, Davis RC, Rodriguez CAG, Vishart A, Deng W, et al. Mechanistic insights into aldehyde production from electrochemical CO_2 reduction on CuAg alloy *via operando* X-ray measurements. ACS Catalysis. 2023;**13**(14):9379-9391. DOI: 10.1021/acscatal.3c01009

[30] Matsusada Precision Inc. Sputtering | Power Supply Application. Matsusada Precision; 2024 [Internet]. Available from: https:// www.matsusada.com/application/ps/ sputtering/ [Accessed: June 14, 2024]

[31] Jeng EK, Zhang Q, Kashi A, Hunegnaw S, Huo Z, Miller J, et al. Scalable gas diffusion electrode fabrication for electrochemical CO_2 reduction using physical vapor deposition methods. ACS Applied Materials and Interfaces. 2022;**14**(6):7731-7740. DOI: 10.1021/acsami.1c17860

[32] Thornton JA. Sputter coating-its principles and potential. SAE Technical Paper Series. 1973;**1**(1):730544. DOI: 10.4271/730544

[33] Perron A, Politano O, Vignal V. Grain size, stress and surface roughness. Surface and Interface Analysis. 2008;**40**(3-4):518-521. DOI: 10.1002/sia.2849

[34] Jiang K, Huang Y, Zeng G, Toma FM, Goddard WA, Bell AT. Effects of surface roughness on the electrochemical reduction of CO_2 over Cu. ACS Energy Letters. 2020;**5**(4):1206-1214. DOI: 10.1021/acsenergylett.0c00482

[35] Wiranarongkorn K, Eamsiri K, Chen YS, Arpornwichanop A. A comprehensive review of electrochemical reduction of CO2 to methanol: Technical and design aspects. Journal of CO2 Utilization. 2023;**71**:102477. DOI: 10.1016/j.jcou.2023.102477

Section 3

Electrocatalysis for Clean Energy Applications

Chapter 5

Perspective Chapter: Beyond the Beaker – The Reality of Splitting Water into a Sustainable Energy Source

Abubakar Tahir Isa, Aliyu Tahir Isah,
Hauwa Mohammed Mari and A.A. Musa

Abstract

Water splitting is a fundamental electrochemical process that holds immense promise for sustainable energy production, primarily through the generation of hydrogen. By using renewable energy sources, water can be split into its constituent elements of hydrogen and oxygen *via* techniques like photocatalysis and electrolysis. The challenge lies in improving the efficiency, scalability, and cost-effectiveness of this process. Advanced materials, such as novel catalysts and semiconductors, are central to optimizing the reaction kinetics and overcoming energy barriers. Recent developments in nano-structured materials, along with advances in photoelectrochemical (PEC) and thermo-chemical water splitting, offer a potential solution to realizing a hydrogen economy.

Keywords: water splitting, catalysis, hydrogen, electrons, holes, tandem catalysis, sustainable energy

1. Introduction

The world is facing a bottleneck in the energy supply, and the growing energy crisis is driven by increasing demand, rapid industrialization, and a rising global population. At the same time, the environmental impact of fossil fuel use, especially in terms of greenhouse gas emissions and climate change, has raised serious concerns [1–3]. These challenges have sparked a worldwide search for cleaner, safer, and more sustainable energy sources. Among the many options being explored, hydrogen stands out as a promising alternative [4]. It is clean, abundant, and efficient, and when used as a fuel, it produces only water as a byproduct. Because of these qualities, hydrogen is now seen as a potential replacement for fossil fuels in the long run [5].

While hydrogen fuel can be generated from methane reforming and other means, which are not friendly to the ecology, one of the most appealing ways to produce hydrogen is by splitting water, a process that uses water molecules to generate hydrogen and oxygen [4]. A number of semiconductor materials have been studied for this application, including metal oxides like MoO_3 and TiO_2, metal sulfides such as CdS, carbides and nitrides like Ti_3C_2, organic compounds like graphitic carbon nitride (g-C_3N_4), and chalcogenides such as MoS_2. These materials have been evaluated both on their own and in combination with others to improve their efficiency in the hydrogen evolution reaction (HER) [6]. Water splitting can be carried out through several techniques, including photocatalysis, electrocatalysis, photoelectrochemical methods, thermochemical processes, and others [7]. Each of these approaches has its own advantages and limitations, and researchers around the globe are working around the clock to improve their efficiency, reduce their costs, and scale them up for practical/industrial use.

This chapter focuses on the principles and progress in water splitting, emphasizing photocatalysis and electrolysis. It discusses major challenges related to efficiency and scale-up as a means to improving performance and enabling practical applications.

2. Water splitting

Water splitting is a fundamental electrochemical process that breaks water down into its basic elements of hydrogen and oxygen. It is gaining significant attention as a hotspot toward clean and sustainable energy, with hydrogen emerging as a promising energy carrier due to its ability to be stored, transported, and used without generating carbon emissions [5, 7]. When powered by renewable energy sources, water splitting offers a way to produce hydrogen independently of fossil fuels. However, this process is not spontaneous and requires a considerable energy input of about $\Delta G = 237$ kJ/mol, or roughly 1.23 eV of electrical potential. This energy drives the excitation of electrons and holes in electrodes or semiconductor materials, which is necessary to initiate and sustain the reaction while minimizing the recombination of charge carriers [8].

Hydrogen and oxygen can be produced from water using a range of energy sources, both renewable and nonrenewable. Nonrenewable approaches include high-temperature thermal splitting at over 2500 K and electrolysis powered by electricity generated from fossil fuels [9]. Despite certain technical hurdles and efficiency concerns, water splitting is widely viewed as a cleaner and more sustainable alternative to traditional hydrogen production methods like steam methane reforming (SMR) [10].

2.1 Electrolytic water splitting

Electrolysis of water is widely regarded as a promising method for hydrogen generation due to its environmentally friendly nature and the potential to utilize electricity resulting from renewable energy sources instead of fossil fuels [11]. Among various water-splitting techniques, electrolysis generally offers higher efficiency and hydrogen yield. However, it is an energy-demanding process that requires substantial power input. The overall energy costs and potential environmental impacts can be significantly mitigated if the electricity is sourced from renewable technologies such as solar, wind, or hydropower, either independently or in combination. A typical electrolyzer comprises several essential components: electrodes (anode and cathode), an electrolyte, a membrane or separator, and a power supply. The electrodes must be conductive, chemically stable, and catalytically active. The electrolyte serves as the medium for ionic transport between electrodes and can be acidic, alkaline, or neutral depending on the system configuration. A membrane or separator is positioned

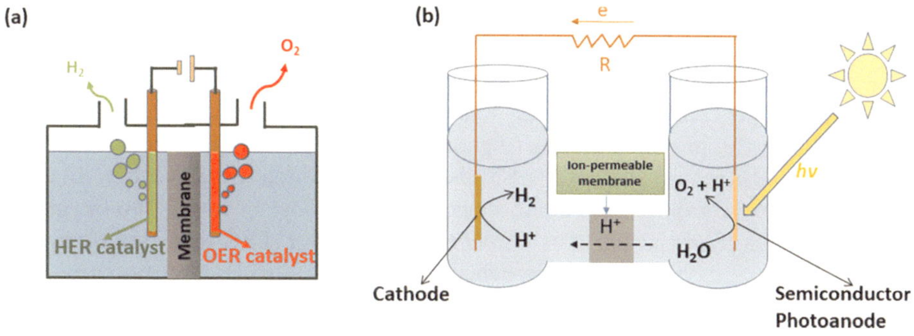

Figure 1.
(a) Electrolytic water splitting and (b) photoelectrolytic water-splitting system.

between the electrodes to prevent gas crossover, ensuring that hydrogen and oxygen gases are kept apart, which is important for both safety and the purity of the product gases [12]. This process is conducted within a system called an electrolyzer (**Figure 1a**), which houses both cathode and anode electrodes where the hydrogen evolution reaction (HER) and oxygen evolution reaction (OER) occur, respectively. The setup also includes a polymer electrolyte membrane (PEM) that facilitates the transfer of protons (hydrogen ions) to the cathode side and contains the aqueous medium (water) [13].

Depending on whether the system operates under acidic or alkaline conditions, the corresponding half-cell reactions for water splitting are expressed in Eqs. (1)–(6). This method is reliable but comes with a few issues. It uses a lot of electricity, which makes it expensive unless powered by cheap renewables. The equipment, especially the catalysts and membranes, is costly and can wear out over time. It also needs very pure water to work properly, and if hydrogen and oxygen mix due to leaks or membrane failure, it can be dangerous.

In acidic solution:

$$\text{Anode}: \ 2H_2O \rightarrow O_2 + 4H^+ + 4e^-; \quad E^0 = -1.23 \text{ eV} \tag{1}$$

$$\text{Cathode}: 4H^+ + 4e^- \rightarrow 2H_2; \quad E^0 = -0.00 \text{ eV} \tag{2}$$

Overall splitting:

$$2H_2O \rightarrow 2O_2 + 2H_2; \quad E^0 = -1.23 \text{ eV} \tag{3}$$

In alkaline solution:

$$\text{Anode}: 4OH^- \rightarrow O_2 + 2H_2O + 4e^-; \quad E^0 = -0.40 \text{ eV} \tag{4}$$

$$\text{Cathode}: \ 4H_2O + 4e^- \rightarrow 2H_2 + 4OH^-; \quad E^0 = -0.83 \text{ eV} \tag{5}$$

Overall splitting:

$$2H_2O \rightarrow 2O_2 + 2H_2; \qquad E^0 = -1.23 \text{ eV} \tag{6}$$

The required thermodynamic potential to carry out water oxidation and reduction, respectively, is $E^0 = -1.23$ eV [14].

2.2 Photoelectrochemical (PEC) water splitting

This approach represents a shift from using fossil-fuel-generated electricity, which powers electrolytic cells, to harnessing solar energy through photovoltaic (PV) cells or, more efficiently, photoelectrodes to generate the electricity needed for water splitting. Since the pioneering work of Fujishima and Honda in 1972, where titanium dioxide (TiO_2) was employed as a photoelectrode to dissociate water into oxygen and hydrogen using sunlight rather than direct electrical input, PEC water splitting has gained attention as a cost-effective hydrogen and oxygen production method [15]. Compared to steam methane reforming (SMR) and conventional electrolysis, PEC offers a greener and more promising pathway to convert solar energy into storable and transportable chemical energy in the form of hydrogen [16]. Similar to photocatalysis, PEC utilizes photocatalytic materials typically applied as thin films or coatings on electrodes to absorb sunlight and drive the oxidation and reduction half-reactions necessary for overall water decomposition (**Figure 1b**) [17]. Notably, integrating photovoltaic cells with PEC systems has led to significant improvements in solar-to-hydrogen (STH) conversion efficiency, with PV-assisted PEC systems achieving efficiencies of approximately 19%, a substantial increase compared to the ~3% observed in standalone PEC setups. However, this technique is still experimental and not ready for large-scale use. It produces very little hydrogen because the materials used are not efficient at capturing sunlight and tend to break down quickly in water. It also only works when the sun is out, which makes it hard to rely on. Overall, it is a good idea on paper but not practical yet [10].

2.3 Photobiological water splitting

Photobiological water splitting, also referred to as photosynthetic biohydrogen production, is a biological method that harnesses solar energy and water in the presence of light-responsive microorganisms to generate hydrogen gas (H_2) [18]. A key advantage of this process is that it operates under ambient temperature and pressure conditions, as the microorganisms involved naturally carry out hydrogen production without the need for extreme environments. These organisms possess photosynthetic systems, along with oxygen-sensitive enzymes such as hydrogenase and nitrogenase, which facilitate the process [19].

A specific example of this approach is *direct photolysis*, which employs green algae or cyanobacteria to drive the water-splitting reaction. In this pathway, water is decomposed into oxygen and hydrogen through the action of hydrogenase enzymes, as represented by Eq. (7). Other biological methods include *photo-fermentation*, *dark fermentation*, *indirect photolysis*, and *CO-gas fermentation* [20].

$$2H_2O + \text{solar energy} \rightarrow 2H_2 + O_2 \quad (7)$$

$$12H_2O + 6CO_2 + \text{solar energy} \rightarrow C_6H_{12}O_6 + 6O_2 \quad (8)$$

$$C_6H_{12}O_6 + 12H_2O + \text{solar energy} \rightarrow + 12H_2 + 6O_2 \quad (9)$$

In indirect photolysis, hydrogen production occurs through a two-stage mechanism: initially, carbon dioxide is fixed and oxygen is generated; this is followed by a separate stage where oxygen evolution and hydrogen production take place, typically described by Eqs. (8) and (9) [11]. Photo-fermentation represents another biophotolytic route for hydrogen generation. This process utilizes light and photosynthetic microorganisms, specifically purple non-sulfur (PNS) bacteria, alongside organic electron donors such as acetone, butyric acid, or lactic acid, to enhance hydrogen yield [21]. Notably, PNS bacteria lack photosystem II (PS II), which distinguishes them from green algae and cyanobacteria and influences their hydrogen production pathway [22–24]. Overall, photobiological hydrogen production is considered an environmentally friendly and cost-effective strategy due to the natural abundance of microalgae and the low-energy input required. Additionally, it contributes no carbon emissions. Nevertheless, several challenges limit its large-scale application, including the low hydrogen yield from photosynthetic microorganisms and the high cost associated with the development and maintenance of photobioreactors [25, 26].

2.4 Photocatalytic water splitting

As discussed earlier, the decomposition of water into oxygen and hydrogen is a non-spontaneous process that necessitates an external energy input, typically in the form of light, electricity, or heat, to proceed [27]. This requirement has led to the development of an alternative hydrogen production method that relies exclusively on sunlight and a particulate semiconductor catalyst to drive the water-splitting reaction. This technique, often referred to as artificial photosynthesis due to its similarity to natural photosynthesis (where chlorophyll acts as the light-absorbing catalyst), allows for the direct transformation of solar energy into chemical energy, thereby eliminating the need for electrical intermediates, as seen in electrolysis and photoelectrochemical (PEC) systems [28, 29].

Although this approach offers the benefit of bypassing electricity, its hydrogen output remains lower than that of electrochemical methods, which are already industrially established, primarily due to the relatively low solar-to-hydrogen (STH) conversion efficiency. Photocatalytic efficiency is typically quantified by the hydrogen generation rate, expressed as moles of hydrogen produced per gram of catalyst per hour (mol/g/h) [30]. Like other water-splitting techniques, photocatalysis is an energy-intensive (endergonic) process, requiring a minimum Gibbs free energy of 237 kJ/mol, equivalent to an electrical potential of 1.23 eV, to drive water decomposition [31]. The overall photocatalytic process involves two key half-reactions occurring at the valence band (VB) and conduction band (CB) of the semiconductor material, with the net reaction described by Eqs. (10)–(12).

$$\text{Photooxidation (CB)} : H_2O + H^+ \rightarrow \frac{1}{2}O_2 + 2H^+; \quad E^0 = -1.23\ \text{eV} \tag{10}$$

$$\text{Photoreduction (VB)} : 2H^+ + 2e^- \rightarrow H_2; \quad E^0 = -0.00\ \text{eV} \tag{11}$$

Overall splitting:

$$H_2O \rightarrow \frac{1}{2}O_2 + 2H_2; \quad E^0 = -1.23\ \text{eV} \tag{12}$$

The successful execution of photocatalytic water splitting depends on several essential criteria:

i. The photocatalyst must possess a conduction band (CB) minimum or, equivalently, the highest occupied molecular orbital (HOMO) that lies at a more positive potential than the redox level for water oxidation (H^+/H_2), which is 1.23 eV versus the normal hydrogen electrode (NHE) at pH 0. Simultaneously, its valence band (VB) maximum or the lowest unoccupied molecular orbital (LUMO) has to be positioned more negatively than the water-reduction potential (H_2O/O_2), which is 0 V versus NHE at the same pH [32].

ii. The energy of incoming photons must exceed the semiconductor's bandgap energy (E_g), meaning *hv* > E_g, e.g., to initiate photoexcitation (**Figure 2**).

iii. For effective light absorption within the visible range of the solar spectrum, the semiconductor's bandgap should ideally fall between 1.23 and 3.15 eV. A bandgap range of 2.0–2.5 eV is generally considered optimal for balancing absorption efficiency and redox potential alignment [27].

2.4.1 Hydrogen production mechanism via *photocatalytic water splitting*

Hydrogen generation through photocatalytic water splitting operates *via* two primary photoexcitation pathways: single-step and two-step excitation mechanisms. In the single-step system, both the oxidation and reduction half-reactions occur simultaneously on the surface of a single semiconductor photocatalyst. In contrast, the two-step excitation approach, often referred to as a Z-scheme, utilizes a heterojunction composed of two different semiconductors, with each half-reaction (oxidation or reduction) taking place on separate materials [33, 34].

Regardless of the excitation scheme employed, the photocatalytic water-splitting process typically involves the following key stages:

1. *Photon absorption and charge generation:* The photocatalyst absorbs incident photons with energy equal to or greater than its bandgap, resulting in the formation of electron-hole pairs.

2. *Charge carrier separation and transfer:* In this stage, the photogenerated electrons transition to the conduction band while the holes remain in the valence band.

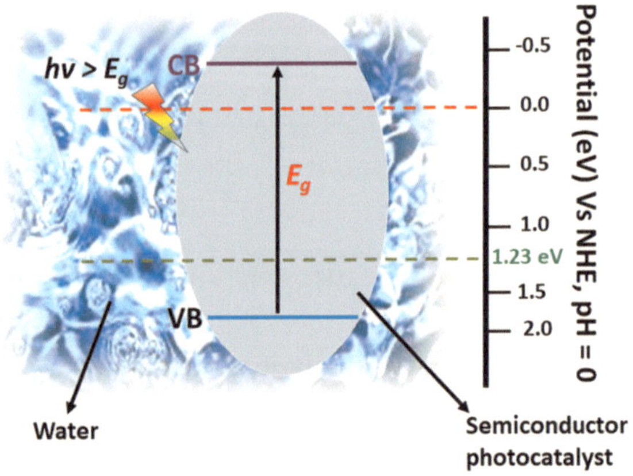

Figure 2.
E_g, CB, and VB requirements for photocatalytic water splitting.

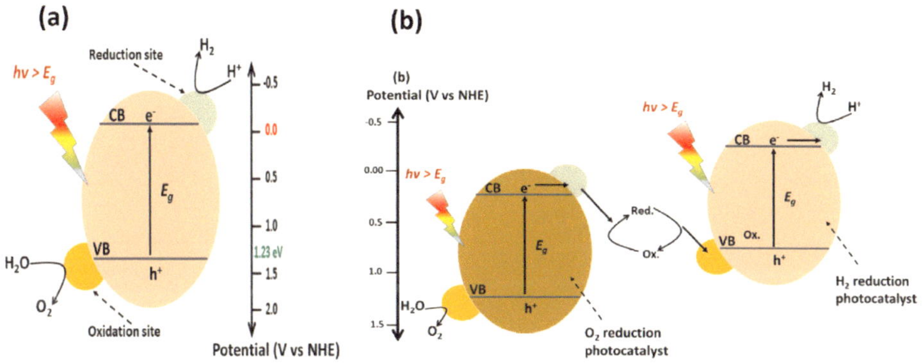

Figure 3.
Photoexcitation (a) one-step and (b) two-step for H_2 evolution using photocatalytic water splitting.

Effective separation and migration of these charge carriers to the photocatalyst surface is essential. In two-step excitation systems, this involves charge exchange between two band-aligned semiconductors; electrons move toward one semiconductor, while holes transfer to the other (**Figure 3**).

3. *Surface redox reactions:* At the photocatalyst's surface, the separated charges drive the redox reactions; electrons reduce protons to generate hydrogen (H_2), and holes oxidize water to produce oxygen (O_2) [31, 35–37].

Each of these steps is pivotal in determining the overall efficiency of the water-splitting process. Therefore, the effectiveness of a photocatalyst is closely tied to its ability to harvest visible light efficiently, facilitate rapid charge separation and transport, and offer suitable surface characteristics for effective adsorption and reaction of the reactant molecules [31].

3. Thermodynamics and kinetics of PWS

3.1 Thermodynamics of water splitting

For water splitting to occur efficiently, the photocatalyst must have a narrow bandgap that can absorb visible light and provide active surface sites for the reaction. Thermodynamically, the energy of the incoming photons (hv) needs to exceed the bandgap energy (E_g) of the semiconductor ($hv > E_g$) to excite electrons and trigger spontaneous electron transfer [38]. This excitation raises the chemical potential of the electrons above that of the holes, resulting in a negative Gibbs free energy change. Such a negative value favors the forward reaction, allowing the excited holes and electrons to participate in surface reactions that produce hydrogen [34].

However, if the reaction reverses, holes and electrons recombine, reducing the hydrogen generation efficiency of the system. Splitting water is particularly challenging because of the strong O—H bonds, requiring about 237 kJ/mol of energy [39]. In contrast, the recombination of hydrogen and oxygen to form water is an exothermic process, meaning it releases energy and is therefore thermodynamically more favorable (**Figure 4**). The water-splitting reaction is uphill energetically and demands a significant input of Gibbs free energy to split 1 mol of water (Eq. (15)), while the reverse reaction proceeds more easily because it is exothermic [31].

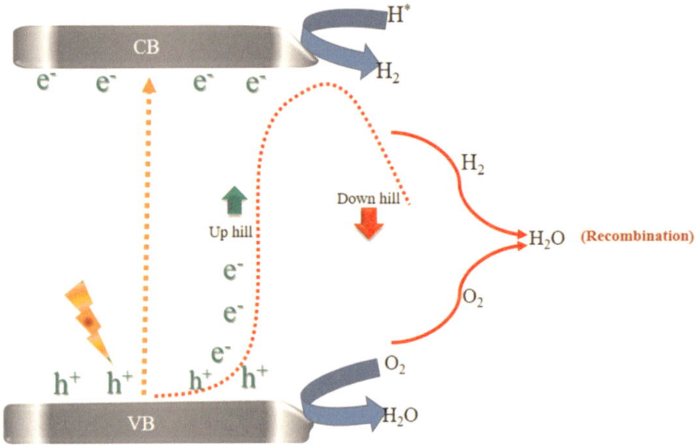

Figure 4.
Kinetics of water splitting.

$$2H^{+}+2e^{-}\rightarrow H_2\ E_{redox}=0\ eV \quad (13)$$

$$2H_2O+4h^{+}\rightarrow 4H^{+}+O_2\ E_{redox}=1.23\ eV \quad (14)$$

$$H_2O\rightarrow H_2+\frac{1}{2}O_2, \Delta G=237\ kJ/mol \quad (15)$$

3.2 Photocatalytic kinetics

This describes how quickly the various steps of a photocatalytic reaction take place, starting from the generation of charge carriers, through their separation and movement, and finally to the reactions occurring on the surface of the catalyst. Due to the fact that hydrogen atoms are about 16 times lighter than oxygen atoms, oxygen molecules tend to diffuse more slowly and have a harder time leaving the photocatalyst surface [40]. Additionally, the separation and migration of electrons happen incredibly fast, typically within femtoseconds to microseconds (10^{-15}–10^{-6} seconds). In contrast, the surface reactions in a solid-liquid environment occur over a slightly longer timescale: microseconds to milliseconds (10^{-6}–10^{-3} seconds). This timing mismatch means that the coulombic attraction between electrons and holes can easily cause them to recombine before contributing to hydrogen production, which limits the overall efficiency [15]. Electron transfer itself takes place within femtoseconds to picoseconds [41], and gaining a deeper understanding of these rapid kinetics is key to designing photocatalysts with better performance [39]. For instance, Zhang and colleagues used femtosecond transient absorption spectroscopy to demonstrate that the electron transfer rate in a CdS-Pt photocatalyst increased from roughly $3.5 \times 10^{10}\ s^{-1}$ to about $4.9 \times 10^{10}\ s^{-1}$ [42]. To tackle the challenge of rapid charge recombination, several approaches have been explored, including controlling the morphology of the photocatalyst and loading it with cocatalysts.

4. Challenges in water splitting

Water splitting, while pivotal for green hydrogen production, confronts formidable scientific and engineering obstacles that limit its scalability and commercial viability. The oxygen evolution reaction (OER), a four-proton-coupled electron transfer process, imposes steep thermodynamic (1.23 V theoretical minimum) and kinetic overpotentials, consuming up to 70% of the total energy input in conventional electrolysis [24]. Noble metal catalysts (e.g., Ir, Ru) remain irreplaceable for efficient OER but face scarcity, cost volatility, and dissolution under harsh operating conditions, while non-precious alternatives struggle with poor conductivity, surface passivation, or instability in acidic/alkaline media. Photoelectrochemical (PEC) systems grapple with material bottlenecks: semiconductors like $BiVO_4$ or Fe_2O_3 suffer from rapid charge recombination, insufficient light absorption, and photocorrosion, whereas tandem designs introduce interfacial compatibility issues [14]. Gas evolution at electrodes creates resistive bubbles, blocking active sites and increasing ohmic losses, while membrane degradation in electrolyzers exacerbates crossover and safety risks. System-level challenges persist, including intermittent renewable energy integration, high capital costs for proton-exchange membrane setups, and the lack of standardized durability metrics for industrial deployment. These intertwined issues underscore the gap between lab-scale breakthroughs and gigawatt-scale hydrogen economies [10].

5. Modification strategies

5.1 Doping

Doping is the process of adding small amounts of another element to a material with a vision to improve its performance. In water splitting, this helps catalysts work better by making reactions faster, boosting efficiency, and increasing stability. For example, adding metals like nickel or iron to a catalyst can help it conduct electricity better, while nonmetals like nitrogen can improve how it absorbs light. The goal is to make water splitting more effective for producing clean hydrogen energy [43–45].

5.1.1 Metal doping

Most of the photocatalytic performance of semiconductors is limited by their edge-confined active sites and inherently low electrical conductivity. To address these drawbacks and improve its hydrogen evolution activity, various strategies have been explored. Among them, metal doping has emerged as a promising approach. Introducing transition metals such as copper (Cu), cobalt (Co), iron (Fe), and silver (Ag) into the MoS_2 structure has been shown to significantly enhance its catalytic efficiency [43, 45]. For example, Zhou et al. synthesized a composite of amorphous MoS_2 doped with cobalt and combined with graphene through a solvothermal method. This hybrid material achieved a hydrogen production rate of 11.45 mmol h^{-1} g^{-1}, representing a 21-fold increase compared to MoS_x/graphene alone. The improvement was attributed to the synergistic effects of cobalt doping, which enhanced the intrinsic catalytic activity of amorphous MoS_2 and the graphene [46].

5.1.2 Nonmetal doping

In addition to transition metal doping, the incorporation of nonmetallic elements such as phosphorus (P) and nitrogen (N) into the semiconductor lattice has also been shown to significantly enhance its photocatalytic hydrogen evolution activity. These nonmetal dopants can increase the density of active sites and effectively narrow the bandgap of photocatalysts, enabling broader absorption of visible light [47].

5.2 Heterojunction

Adjusting the charge transfer pathways through strategic band alignment between two photocatalysts has emerged as a vital approach for enhancing hydrogen evolution efficiency in photocatalytic water splitting [48]. The relative positioning of the conduction and valence bands in such binary systems leads to the formation of distinct heterojunction types (**Figure 5**) [38]: *Type I* (straddling gap), where the band edges of one semiconductor lie entirely within the bandgap of the other; *Type II* (staggered gap), characterized by spatial separation of electrons and holes due to offset band edges; and *Type III* (broken gap), where the valence band of one semiconductor (SC I) lies above the conduction band of the other (SC II), resulting in a discontinuous energy overlap [49, 50].

Despite their widespread use, these conventional heterojunctions often face limitations. In particular, they may fail to satisfy the thermodynamic requirements for water splitting if the conduction band of SC I is not sufficiently negative to drive proton reduction or if the valence band of SC II is not sufficiently positive to enable water oxidation [51].

To overcome these limitations, advanced heterojunction architectures such as the *Z-scheme* and *S-scheme* have been proposed (**Figure 6**). The Z-scheme system employs an internal or external charge transfer mediator to facilitate directional electron flow between SC I and SC II, mimicking the natural photosynthetic process. In contrast, the S-scheme heterojunction leverages band bending and internal electric fields at the interface to promote selective recombination of low-energy charge carriers while preserving high-energy electrons and holes for redox reactions [8].

5.3 Loading a suitable cocatalyst

Adding a cocatalyst to the surface of photocatalysts helps trap the charge carriers produced from photoexcitation. This improves how well these charges move, separate, and transfer within the material. Sometimes, the cocatalyst also acts as a reaction site, lowering the energy needed for the hydrogen evolution reaction (HER) [52].

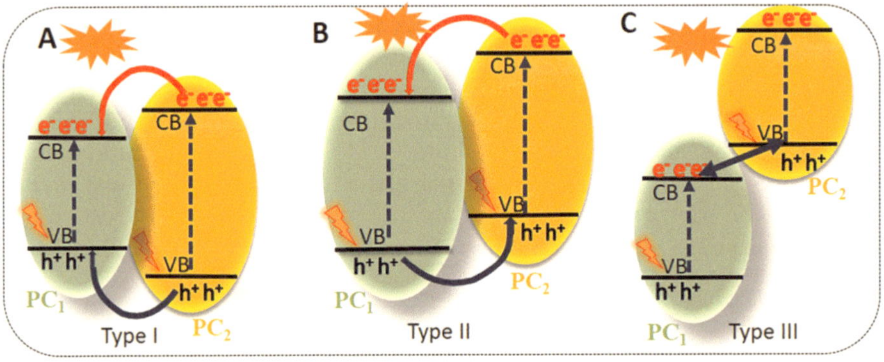

Figure 5.
Band alignment for (A) type I, (B) type II, and (C) type III.

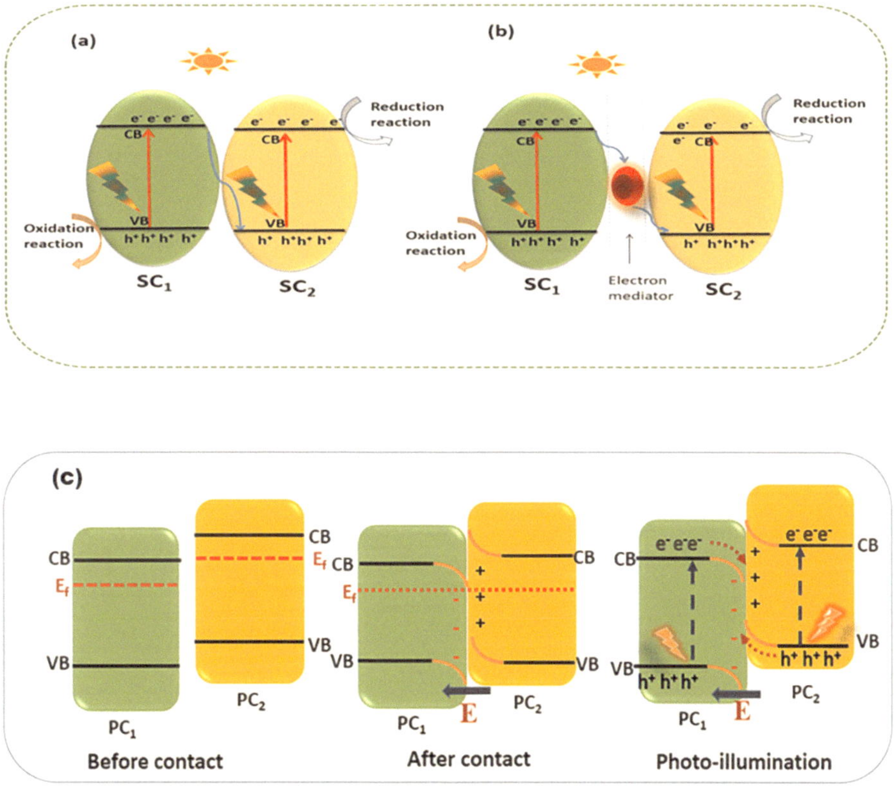

Figure 6.
Schematic illustration of (a) direct Z-scheme, (b) indirect Z-scheme heterojunctions, and (c) pictorial representation of S-scheme heterojunctions.

Researchers have found that certain materials work well as cocatalysts when combined with others. This is because when a single photocatalyst is used to single-handedly split water, it suffers ultrafast charge carrier recombination, which is the general problem with overall water splitting, apart from its inherent limitation. It needs to pair up with other materials so that it will benefit from the synergy, mostly by tuning the bandgap, improving light absorption, etc. These materials include transition metal sulfides like selenides and tellurides [53]. Others are transition metal carbides and oxides, such as MXenes and titanium oxide, which reduce charge carrier recombination and improve the stability of the material under the reaction state [54].

Although they are expensive, noble metals like platinum (Pt), gold (Au), palladium (Pd), ruthenium (Ru), and rhodium (Rh) are also very effective cocatalysts. To perform well in HER, a good cocatalyst should be able to pull electrons away from the photocatalyst and transfer them efficiently to its surface. This requires the right band alignment and excellent electrical conductivity [55–57]. **Table 1** presents some of the photocatalysts in use and their performance in hydrogen generation.

5.4 Role of sacrificial agent

To improve the photocatalytic performance of a material (composite or catalyst), researchers often add organics like glycerol, ethanol, lactic acid, or methanol to water during testing. These substances, called sacrificial agents, help improve the activity of

Photocatalyst	Sacrificial agent	Light source	Synthesis	H_2 production (μmol/g/h)	Ref.
MoS_2/CdS	Na_2SO_4	420 W Xe lamp	Hydrothermal	550	[58]
$LaFeO_3$	Ethanol	400 W Xe lamp	Sol-gel	3315	[59]
$LaNiO_3$-rGO	Methanol	400 W Xe lamp	Facile self-assembly	3220	[60]
$LaFeO_3$/CdS/CQDs	Lactic acid	Sunlight	Hydrothermal	25,302	[61]
$SnNb_2O_6/MoS_2$	Methanol	300 W Xe lamp	Hydrothermal	12.9	[62]
MoS_2@ZnO	Na_2SO_3/Na_2S	150 W Xe lamp	Electrospark erosion	906.6	[63]
$TiO_2@MoS_2$	Ethanol	300 W Xe lamp	Hydrothermal	2720	[64]
$MoS_2@TiO_2@Ti_3C_2$	TEOA	300 W Xe lamp	Hydrothermal	10,505.8	[65]
$MoS_2/TiO_2/NiO_x$	Methanol	300 W Xe lamp	Lithium salt-assisted chemical exfoliation	1629.1	[66]
MoS_2/ZnO	Na_2SO_3/Na_2S	300 W Xe lamp	Hydrothermal	235	[67]
ZnO/MoS_2	Na_2SO_3/Na_2S	300 W Xe lamp	Hydrothermal	165	[68]
MoS_2/N-ZnO	Na_2SO_3/Na_2S	Sunlight	Hydrothermal	17,300	[69]
Cu_2O/CuO	Methanol	300 W Xe lamp	Hydrothermal	981.8	[70]
CuO/Co_2O_4	Ethanol	300 W Xe lamp	Hydrothermal	2143.2	[71]
CuO/$CuP_{12}M_3$	TEOA	300 W Xe lamp	Hydrothermal	1796.8	[72]
$Zn_{0.7}Cd_{0.3}S/Fe_2O_3$	Na_2SO_3, Na_2S	300 W ($\lambda \geq 420$ nm)	Co-precipitation	2724	[73]
ZIF-67@CdS	LA	300 W ($\lambda \geq 420$ nm)	Solvothermal	2744.12	[74]
FeCdS-CdS	Na_2SO_3, Na_2S	300 W ($\lambda \geq 420$ nm)	Thermal vulcanization	7956	[75]
C-doped CdS@G	Na_2SO_3, Na_2S	300 W ($\lambda \geq 420$ nm)	Hydrothermal	3120	[76]
CdS/CoP	Na_2S/Na_2SO_3	300 W ($\lambda \geq 420$ nm)	Precipitation, vulcanization, and sulfidation	15,740	[77]
CdS/np-rGO	Na_2S/Na_2SO_3	300 W ($\lambda \geq 420$ nm)	Homothermal chemical bath deposition	2171.23	[78]

Table 1.
Comparison of H_2 production with different semiconductor materials.

the photocatalyst by capturing the positive charge carriers (holes) or by getting oxidized themselves. This process supports the reduction of hydrogen ions into hydrogen gas (H_2) by the photocatalyst. Because of this, scientists are actively searching for more effective compounds to enhance hydrogen production through water splitting [29].

5.5 Nanostructuring

The strong link between the morphology of semiconductor photocatalysts and their efficiency in hydrogen production has been well established. Key factors such as charge transfer kinetics, the presence of active crystal facets, and light absorption, all crucial to the rate of H_2 generation, are heavily influenced by the size and shape of the photocatalyst. Many researchers have dedicated considerable effort to identifying the optimal morphology and structure of catalysts in hydrogen production [6, 79]. Various morphologies have been fabricated and studied, including zero-dimensional (0D) nanoparticles, one-dimensional (1D) nanorods and nanowires, two-dimensional (2D) nanosheets, and three-dimensional (3D) forms like nanotrees and nanoballs. These morphologies were examined not only for their individual hydrogen production efficiencies but also for their synergistic interactions with different cocatalysts [44]. It is an important way to improve how well they work for water splitting. By shaping materials into tiny forms like nanoparticles, nanorods, or thin sheets, scientists can increase the surface area and expose more places where the reaction can happen. This also helps the material catch more light, making the process more efficient. At the same time, these small structures make it easier for the electric charges created by light to move around without bumping into each other and canceling out. This means more charges can reach the surface to help produce hydrogen [37]. The shape and crystal faces of these nanostructures can also be designed to make the catalyst more active. Nanostructuring also allows catalysts to work better with other materials, forming special junctions that help keep charges apart and improve light absorption. Experts usually look for clear explanations of how these tiny structures are made using methods like hydrothermal or solvothermal and how their size and shape are checked with different tools. It is also important to show how these changes lead to better hydrogen production, backed up by tests that measure the reaction rates and efficiency [5, 7, 80].

5.6 Tandem catalysis

Tandem catalysis is an astute way to improve the water-splitting efficiency by using two different catalysts, each doing what it does best. You see, splitting water into oxygen and hydrogen involves two separate reactions: one to produce hydrogen (called the hydrogen evolution reaction) and another to produce oxygen (the oxygen evolution reaction). Usually, these reactions need different conditions and catalysts to work well, so trying to do both perfectly in one system can be tricky. Tandem catalysis solves this by pairing catalysts that specialize in each half of the reaction, allowing both processes to happen smoothly side by side [81]. The idea is to keep these two reactions either separated in space or time, which helps avoid energy losses and stops the charges from simply recombining before they can do useful work. For example, in some photoelectrochemical cells, one material acts as a photoanode to handle oxygen production, while another material works as a photocathode to generate hydrogen. Because these materials can be chosen to absorb different parts of sunlight, tandem systems can make better use of the solar spectrum than a single catalyst alone, resulting in higher overall efficiency. Sometimes, tandem catalysis goes beyond just light-driven reactions. It can also combine different ways to drive the process, like using light to excite electrons in one catalyst and an external electrical bias to push the reaction forward in another. This teamwork helps lower the energy barrier for splitting water, making it more practical for real-world hydrogen production [20, 81, 82].

Of course, to make tandem systems work well, the interface between the catalysts needs to be carefully designed so electrons and holes can move smoothly without getting trapped or lost. Stability is also key; these materials have to withstand harsh conditions without breaking down. That is why researchers spend a lot of time fine-tuning the materials and their band alignments and adding cocatalysts to improve performance. They use advanced tools to study how charges move and how efficient the overall system is, which helps them improve the design even further [48, 79, 83].

6. Conclusion

Water splitting stands out as a vital process for clean hydrogen production, offering a sustainable way to meet future energy demands. Here, various water-splitting methods are explored, including photocatalysis, electrolysis, and biophotolysis, each with its own strengths and challenges. What is clear is that no single approach is perfect yet, but all benefit greatly from improvements in materials and system design. One of the main takeaways is how modification strategies such as doping with metals or nonmetals, forming heterojunctions, and engineering band alignments play a crucial role in overcoming the inherent limitations of pure photocatalysts. These strategies help increase active sites, improve charge separation, and extend light absorption, which directly improve hydrogen evolution efficiency.

Moreover, advanced nanostructured materials and novel composite systems have shown promising results in enhancing catalytic performance. The development of Z-scheme and S-scheme heterojunctions, for example, highlights how controlling charge transfer routes can dramatically improve the overall activity and stability of photocatalysts. While challenges remain, such as scaling up, cost reduction, and long-term durability, the ongoing innovations in material science and catalyst engineering provide a solid foundation. Continued research in these areas is essential to push water-splitting technologies closer to commercial viability, ultimately enabling a cleaner, hydrogen-based energy economy.

Acknowledgements

We appreciate the effort of the chief Editor and the Reviewers for thoroughly going through the article, and finding the chapter worthy in the scientific community.

Author details

Abubakar Tahir Isa[1*], Aliyu Tahir Isah[2], Hauwa Mohammed Mari[3] and A.A. Musa[4]

1 Federal University Dutse, Jigawa State, Nigeria

2 Yobe State University Damaturu, Nigeria

3 Federal Collage of Education Technical (T), Potiskum, Yobe State, Nigeria

4 Jigawa State Polytechnic, Dutse, Nigeria

*Address all correspondence to: abubakartahirisa@gmail.com

References

[1] Almomani F, Al-Rababah A, Tawalbeh M, Al-Othman A. A comprehensive review of hydrogen generation by water splitting using 2D nanomaterials: Photo vs electro-catalysis. Fuel. 2023;**332**:125905. DOI: 10.1016/j.fuel.2022.125905

[2] Idris MG, Hafeez HY, Mohammed J, Suleiman AB, Ndikilar CE. A review on recent development in the spinel ferrites-based materials for efficient solar fuel (hydrogen) generation via photocatalytic water-splitting. Applied Surface Science Advances. 2023;**18**: 100468-100478. DOI: 10.1016/j.apsadv.2023.100468

[3] Yusuf H, Mohammed J, Ebenezer C, Balarabe A, Salihu R, Muhammad I. Synergistic utilization of magnetic rGO/ $NiFe_2O_4$-g-C_3N_4S-Scheme heterostructure photocatalyst with enhanced charge carrier separation and transfer: A highly stable and robust photocatalyst for efficient solar fuel (hydrogen) generation. Ceramics International. 2023;**49**:5269-5278. DOI: 10.1016/j.ceramint.2022.10.045

[4] Jayabal S, Wu J, Chen J, Geng D, Meng X. Metallic 1T-MoS_2 nanosheets and their composite materials: Preparation, properties and emerging applications. Materials Today Energy. 2018;**10**:264-279. DOI: 10.1016/j.mtener.2018.10.009

[5] Balan B, Xavier MM, Mathew S. MoS_2-based nanocomposites for photocatalytic hydrogen evolution and carbon dioxide reduction. ACS Omega. 2023;**8**:25649-25673. DOI: 10.1021/acsomega.3c02084

[6] Khan MI, Hasan MS, Bhatti KA, Rizvi H, Wahab A, Rehman SU, et al. Effect of Ni doping on the structural, optical and photocatalytic activity of MoS_2, prepared by hydrothermal method. Materials Research Express. 2020;**7**:015061-015071. DOI: 10.1088/2053-1591/ab66f7

[7] Kafadi ADG, Hafeez HY, Mohammed J, Ndikilar CE, Suleiman AB, Isah AT. A recent prospective and progress on MXene-based photocatalysts for efficient solar fuel (hydrogen) generation via photocatalytic water-splitting. International Journal of Hydrogen Energy. 2024;**53**:1242-1258. DOI: 10.1016/j.ijhydene.2023.12.111

[8] Tahir M, Tasleem S, Tahir B. Recent development in band engineering of binary semiconductor materials for solar driven photocatalytic hydrogen production. International Journal of Hydrogen Energy. 2020;**45**:15985-16038. DOI: 10.1016/j.ijhydene.2020.04.071

[9] Isa AT, Hafeez HY, Mohammed J, Ndikilar CE, Suleiman AB, Kafadi ADG. Photocatalytic performance of MXenes co-catalyst in hydrogen (H_2) production via photocatalytic water splitting: A review. Journal of Alloys and Compounds. 2024;**1005**:175951-175961. DOI: 10.1016/j.jallcom.2024.175951

[10] El ouardi M, El Idrissi A, Arab M, Zbair M, Haspel H, Saadi M, et al. Review of photoelectrochemical water splitting: From quantitative approaches to effect of sacrificial agents, oxygen vacancies, thermal and magnetic field on (photo)electrolysis. International Journal of Hydrogen Energy. 2024;**51**:1044-1067. DOI: 10.1016/j.ijhydene.2023.09.111

[11] Tee SY, Win KY, Teo WS, Koh LD, Liu S, Teng CP, et al. Recent progress in energy-driven water splitting. Advancement of Science. 2017;**4**:

1600337-1600347. DOI: 10.1002/advs.201600337

[12] Chakrabarty P, Alam KCA, Paul SK, Saha SC. Hydrogen production by electrolysis: A sustainable pathway. Hydrogen Energy Conversion Management; 2023. pp. 81-102. DOI: 10.1016/B978-0-443-15329-7.00013-2

[13] You B, Sun Y. Innovative strategies for electrocatalytic water splitting. Accounts of Chemical Research. 2018;**51**: 1571-1580. DOI: 10.1021/acs.accounts.8b00002

[14] Li J, Wang S, Chang J, Feng L. A review of Ni based powder catalyst for urea oxidation in assisting water splitting reaction. Advanced Powder Materials. 2022;**1**:100030-100040. DOI: 10.1016/j.apmate.2022.01.003

[15] Villa K, Galán-Mascarós JR, López N, Palomares E. Photocatalytic water splitting: Advantages and challenges. Sustainable Energy & Fuels. 2021;**5**: 4560-4569. DOI: 10.1039/d1se00808k

[16] Ayodhya D. Semiconductors-based Z-scheme materials for photoelectrochemical water splitting: A review. Electrochimica Acta. 2023;**448**: 142118-142128. DOI: 10.1016/j.electacta.2023.142118

[17] Alfa I, Hafeez HY, Mohammed J, Abdu S, Suleiman B, Ndikilar CE. A recent progress and advancement on MoS_2-based photocatalysts for efficient solar fuel (hydrogen) generation via photocatalytic water splitting. International Journal of Hydrogen Energy. 2024;**71**:1006-1025. DOI: 10.1016/j.ijhydene.2024.05.203

[18] Gupta NM. Factors affecting the efficiency of a water splitting photocatalyst: A perspective. Renewable and Sustainable Energy Reviews. 2017; **71**:585-601. DOI: 10.1016/j.rser.2016.12.086

[19] Akhlaghi N, Najafpour-Darzi G. A comprehensive review on biological hydrogen production. International Journal of Hydrogen Energy. 2020;**45**: 22492-22512. DOI: 10.1016/j.ijhydene.2020.06.182

[20] Gautam R, Nayak JK, Ress NV, Steinberger-Wilckens R, Ghosh UK. Bio-hydrogen production through microbial electrolysis cell: Structural components and influencing factors. Chemical Engineering Journal. 2023;**455**:140535-140545. DOI: 10.1016/j.cej.2022.140535

[21] Bora J, Ghosh S, Mitra A. Photobiological hydrogen production by microorganisms. Solar-Driven Green Hydrogen Generation and Storage; 2023. pp. 237-257. DOI: 10.1016/B978-0-323-99580-1.00009-1

[22] Chen J, Li Q, Wang L, Fan C, Liu H. Advances in whole-cell photobiological hydrogen production. Advanced NanoBiomed Research. 2021;**1**:2000051-2000061. DOI: 10.1002/anbr.202000051

[23] Jose V. Photobiological hydrogen as a renewable fuel. The Holistic Approach to Environment. 2021;**11**:67-71. DOI: 10.33765/thate.11.2.4

[24] Mukherjee N, Srivastava R. Photobiological hydrogen production: Introduction and fundamental concept. Solar-Driven Green Hydrogen Generation and Storage; 2023. pp. 193-222. DOI: 10.1016/B978-0-323-99580-1.00012-1

[25] Zhang Q, Liu H, Shui X, Li Y, Zhang Z. Research progress of additives in photobiological hydrogen production system to enhance biohydrogen.

Bioresource Technology. 2022;**362**: 127787. DOI: 10.1016/j.biortech.2022. 127787

[26] Dasgupta CN, Jose Gilbert J, Lindblad P, Heidorn T, Borgvang SA, Skjanes K, et al. Recent trends on the development of photobiological processes and photobioreactors for the improvement of hydrogen production. International Journal of Hydrogen Energy. 2010;**35**:10218-10238. DOI: 10.1016/j.ijhydene.2010.06.029

[27] Suresh R, Gnanasekaran L, Rajendran S, Soto-Moscoso M. Doped nanomaterials: Application in hydrogen production via photocatalytic water splitting. Fuel. 2023;**348**:128528-128538. DOI: 10.1016/j.fuel.2023.128528

[28] Fajrina N, Tahir M. ScienceDirect A critical review in strategies to improve photocatalytic water splitting towards hydrogen production. International Journal of Hydrogen Energy. 2018;**44**: 540-577. DOI: 10.1016/j.ijhydene. 2018.10.200

[29] Qanugo K, Bose D, Thakur KK. A review on photocatalytic water splitting. E3S Web of Conferences. 2021;**309**:1-12. DOI: 10.1051/e3sconf/202130901032

[30] Corredor J, Rivero MJ, Rangel CM, Gloaguen F, Ortiz I. Comprehensive review and future perspectives on the photocatalytic hydrogen production. Journal of Chemical Technology and Biotechnology. 2019;**94**:3049-3063. DOI: 10.1002/jctb.6123

[31] Tentu RD, Basu S. Photocatalytic water splitting for hydrogen production. Current Opinion in Electrochemistry. 2017;**5**:56-62. DOI: 10.1016/j. coelec.2017.10.019

[32] Chong KC, Li C, Liu B. Recent advances in organic photocatalysts for solar water splitting. CCS Chemistry. 2023;**5**:2436-2447. DOI: 10.31635/ ccschem.023.202302868

[33] Kumar D, Abraham JE, Varghese M, George J, Balachandran M, Cherusseri J. Nanocarbon assisted green hydrogen production: Development and recent trends. International Journal of Hydrogen Energy. 2024;**50**:118-141. DOI: 10.1016/j.ijhydene.2023. 07.257

[34] Li Y, Tsang SCE. Recent progress and strategies for enhancing photocatalytic water splitting. Materials Today Sustainability. 2020;**9**:100032-100042. DOI: 10.1016/j.mtsust.2020. 100032

[35] Zhu S, Wang D. Photocatalysis: Basic principles, diverse forms of implementations and emerging scientific opportunities. Advanced Energy Materials. 2017;**7**:1-24. DOI: 10.1002/ aenm.201700841

[36] Yang J, Wang D, Han H, Li C. Roles of cocatalysts in photocatalysis and photoelectrocatalysis. Accounts of Chemical Research. 2013;**46**:1900-1909. DOI: 10.1021/ar300227e

[37] Chen Y, Feng X, Guo X, Zheng W. Toward a fundamental understanding of factors affecting the function of cocatalysts in photocatalytic water splitting. Current Opinion in Green and Sustainable Chemistry. 2019;**17**:21-28. DOI: 10.1016/j.cogsc.2019.01.002

[38] Zhang L, Jin Z, Tsubaki N. Activating and optimizing the MoS_2@MoO_3 S-scheme heterojunction catalyst through interface engineering to form a sulfur-rich surface for photocatalyst hydrogen evolution. Chemical Engineering Journal. 2022;**438**: 135238- 135242. DOI: 10.1016/j. cej.2022.135238

[39] Shandilya P, Sambyal S, Sharma R, Mandyal P, Fang B. Properties, optimized morphologies, and advanced strategies for photocatalytic applications of WO_3 based photocatalysts. Journal of Hazardous Materials. 2022;**428**:128218. DOI: 10.1016/j.jhazmat.2022.128218

[40] Domen K, Takahashi Y, Onishi H. Transient kinetics of O_2 evolution in photocatalytic water-splitting reaction. ACS Catalysis. 2020;**10**: 13159-13164. DOI: 10.1021/acscatal.0c04115

[41] Takanabe K. Photocatalytic water splitting: Quantitative approaches toward photocatalysis by design. ACS Catalysis. 2017;7:8006-8022. DOI: 10.1021/acscatal.7b02662

[42] Zhang J, Yang G, He B, Cheng B, Li Y, Liang G, et al. Electron transfer kinetics in CdS/Pt heterojunction photocatalyst during water splitting. Chinese Journal of Catalysis. 2022;**43**: 2530-2538. DOI: 10.1016/S1872-2067(22)64108-1

[43] Hong S, Kumar DP, Kim EH, Park H, Gopannagari M, Reddy DA, et al. Earth abundant transition metal-doped few-layered MoS_2 nanosheets on CdS nanorods for ultra-efficient photocatalytic hydrogen production. Journal of Materials Chemistry A. 2017;**5**: 20851-20859. DOI: 10.1039/c7ta06556f

[44] Chen J, Kang N, Fan J, Lu C, Lv K. Carbon nitride for photocatalytic water splitting to produce hydrogen and hydrogen peroxide. Materials Today Chemistry. 2022;**26**:101028. DOI: 10.1016/j.mtchem.2022.101028

[45] Li J, Peng Y, Qian X, Lin J. Few-layer Co-doped MoS_2 nanosheets with rich active sites as an efficient cocatalyst for photocatalytic H_2 production over CdS. Applied Surface Science. 2018;**452**: 437-442. DOI: 10.1016/j.apsusc.2018.05.021

[46] Zhou H, Liu Y, Zhang L, Li H, Liu H, Li W. Transition metal-doped amorphous molybdenum sulfide/graphene ternary cocatalysts for excellent photocatalytic hydrogen evolution: Synergistic effect of transition metal and graphene. Journal of Colloid and Interface Science. 2019;**533**:287-296. DOI: 10.1016/j.jcis.2018.07.084

[47] Lee TH, Kim SY, Jang HW. Black phosphorus: Critical review and potential for water splitting photocatalyst. Nanomaterials. 2016;**6**: 194-199. DOI: 10.3390/nano6110194

[48] Isa AT, Hafeez HY, Mohammed J, David A, Kafadi G, Ndikilar CE, et al. A review on the progress and prospect of CdS-based photocatalysts for hydrogen generation via photocatalytic water splitting. Journal of Alloys and Compounds Communications. 2024;**4**: 100043. DOI: 10.1016/j.jacomc.2024.100043

[49] Yan Y, Zeng Z, Huang M, Chen P. Van der waals heterojunctions for catalysis. Materials Today Advances. 2020;**6**:100059-100068. DOI: 10.1016/j.mtadv.2020.100059

[50] Low J, Yu J, Jaroniec M, Wageh S, Al-Ghamdi AA. Heterojunction photocatalysts. Advanced Materials. 2017;**29**:1-20. DOI: 10.1002/adma.201601694

[51] Maleki A, Saberi MH, Moallemi SA, Jazayeri MH. Evaluation of hydrocarbon generation potential of source rock using two-dimensional modeling of sedimentary basin: A case study in North Dezful Embayment, Southwest Iran. Journal of Petroleum Exploration and Production. 2021;**11**:2861-2876. DOI: 10.1007/s13202-021-01202-5

[52] Yu Y, Li W, Yang H, Wei Q, Hou L, Wu Z, et al. 4-Methyl-5-vinyl thiazole modified Ni-MOF/g-C3N4/CdS composites for efficient photocatalytic hydrogen evolution without precious metal cocatalysts. Journal of Colloid and Interface Science. 2023;**651**:221-234. DOI: 10.1016/j.jcis.2023.07.210

[53] Umaru D, Yusuf H, Mohammed J, Balarabe A, Ebenezer C, Zakariyya Y. A review of recent progress in solar fuel (hydrogen) generation via photocatalytic water-splitting of cadmium sulfide (CdS) based photocatalyst. Applied Surface Science Advances. 2023;**18**:100520. DOI: 10.1016/j.apsadv.2023.100520

[54] Zhao N, Hu Y, Du J, Liu G, Dong B, Yang Y, et al. $Ti_3C_2T_x$ MXene-derived amorphous TiO_2-C nanosheet cocatalysts coupled CdS nanostructures for enhanced photocatalytic hydrogen evolution. Applied Surface Science. 2020;**530**:147247. DOI: 10.1016/j.apsusc.2020.147247

[55] Ran J, Gao G, Li F, Ma T, Du A, Qiao S. Metal sulfide photo-absorbers for enhanced visible-light photocatalytic hydrogen production. Nature Communications. 2017;**10**:13907. DOI: 10.1038/ncomms13907

[56] Li K, Zhang S, Li Y, Fan J, Lv K. MXenes as noble-metal-alternative co-catalysts in photocatalysis. Chinese Journal of Catalysis. 2021;**42**:3-14. DOI: 10.1016/S1872-2067(20)63630-0

[57] Solomon G. Engineering inorganic nanostructured composites for boosting H_2 and O_2 evolution reactions. Materials Chemistry Composite Science and Engineering. 2022:18489-18501

[58] Zong X, Wu G, Yan H, Ma G, Shi J, Wen F, et al. Photocatalytic H_2 evolution on MoS_2/CdS catalysts under visible light irradiation. The Journal of Physical Chemistry C. 2010;**114**:1963-1968

[59] Tijare SN, Joshi MV, Padole PS, Mangrulkar PA, Rayalu SS, Labhsetwar NK. Photocatalytic hydrogen generation through water splitting on nano-crystalline $LaFeO_3$ perovskite. International Journal of Hydrogen Energy. 2012;**37**:10451-10456. DOI: 10.1016/j.ijhydene.2012.01.120

[60] Lv T, Wu M, Guo M, Liu Q, Jia L. Self-assembly photocatalytic reduction synthesis of graphene-encapusulated $LaNiO_3$ naanoreactor with high efficiency and stability for photocatalytic water splitting to hydrogen. Chemical Engineering Journal. 2019;**356**:580-591. DOI: 10.1016/j.cej.2018.09.031

[61] Manchala S, Gandamalla A, Rao N. High potential and robust ternary $LaFeO_3$/CdS/carbon quantum dots nanocomposite for photocatalytic H_2 evolution under sunlight illumination. Journal of Colloid and Interface Science. 2021;**583**:255-266. DOI: 10.1016/j.jcis.2020.08.125

[62] Jiang D, Wen B, Zhang Y, Jin Y, Li D, Chen M. MoS_2/$SnNb_2O_6$ 2D/2D nanosheet heterojunctions with enhanced interfacial charge separation for boosting photocatalytic hydrogen evolution. Journal of Colloid and Interface Science. 2019;**536**:1-8. DOI: 10.1016/j.jcis.2018.10.027

[63] An V, Potgieter H, Usoltseva N, Valiev D, Stepanov S, Pustovalov A, et al. MoS_2@ZnO nanoheterostructures prepared by electrospark erosion for photocatalytic applications. Nanomaterials. 2021;**11**:157. DOI: 10.3390/nano11010157

[64] Zhong C, Weng W, Liang X, Gu D, Xiao W. One-step molten-salt synthesis of anatase/rutile bi-phase TiO_2@MoS_2

hierarchical photocatalysts for enhanced solar-driven hydrogen generation. Applied Surface Science. 2020;**507**: 145072. DOI: 10.1016/j.apsusc. 2019.145072

[65] Li Y, Ding L, Liang Z, Xue Y, Cui H, Tian J. Synergetic effect of defects rich MoS_2 and Ti3C2 MXene as cocatalysts for enhanced photocatalytic H_2 production activity of TiO_2. Chemical Engineering Journal. 2020;**383**:123178-123188. DOI: 10.1016/j.cej.2019. 123178

[66] Lin H, Zhang K, Yang G, Li Y, Liu X, Chang K, et al. Ultrafine nano 1T-MoS_2 monolayers with NiOx as dual co-catalysts over TiO_2 photoharvester for efficient photocatalytic hydrogen evolution. Applied Catalysis B: Environmental. 2020;**279**:119387. DOI: 10.1016/j.apcatb.2020. 119387

[67] Hunge YM, Yadav AA, Kang SW, Jun Lim S, Kim H. Visible light activated MoS_2/ZnO composites for photocatalytic degradation of ciprofloxacin antibiotic and hydrogen production. Journal of Photochemistry and Photobiology A: Chemistry. 2023;**434**:1-10. DOI: 10.1016/ j.jphotochem.2022.114250

[68] Tahir MB, Sohaib M, Rafique M, Sagir M, Rehman NU, Muhammad S. Visible light responsive photocatalytic hydrogen evolution using MoS_2 incorporated ZnO. Applied Nanoscience. 2020;**10**:3925-3931. DOI: 10.1007/ s13204-020-01476-x

[69] Kumar S, Kumar A, Navakoteswara Rao V, Kumar A, Shankar MV, Krishnan V. Defect-rich MoS_2 ultrathin nanosheets-coated nitrogen-doped ZnO nanorod heterostructures: An insight into in-situ-generated ZnS for enhanced photocatalytic hydrogen evolution. ACS Applied Energy Materials. 2019;**2**: 5622-5634. DOI: 10.1021/acsaem. 9b00790

[70] Zhang Y, Liu M, Chen J, Xie K, Fang S. Dendritic branching Z-scheme Cu_2O/TiO_2 heterostructure photocatalysts for boosting H_2 production. Journal of Physics and Chemistry of Solids. 2021;**152**. DOI: 10.1016/j.jpcs.2021.109948

[71] Wang L, Huang G, Zhang L, Lian R, Huang J, She H, et al. Construction of TiO_2-covalent organic framework Z-Scheme hybrid through coordination bond for photocatalytic CO_2 conversion. Journal of Energy Chemistry. 2022;**64**: 85-92. DOI: 10.1016/j. jechem.2021.04.053

[72] Chen Y, Wang M, Yan F, Wang W, Qu Y, Fang H, et al. Enhanced UV–vis photoinduced hydrogen evolution of metalloporphyrin sensitized PSf/TiO_2 MMMs by varying center metal ion complexed in porphyrin. Fuel. 2022;**312**: 122810. DOI: 10.1016/j.fuel.2021.122810

[73] Bai J, Shen R, Chen W, Xie J, Zhang P, Jiang Z. Enhanced photocatalytic H_2 evolution based on a $Ti_3C_2/Zn_{0.7}Cd_{0.3}S/Fe_2O_3$ Ohmic/S-scheme hybrid heterojunction with cascade 2D coupling interfaces. Chemical Engineering Journal. 2022;**429**: 132587. DOI: 10.1016/j.cej.2021.132587

[74] Yu Z, Qian L, Zhong T, Ran Q, Huang J, Hou Y, et al. Enhanced visible light photocatalytic activity of CdS through controllable self-assembly compositing with ZIF-67. Molecular Catalysis. 2020;**485**:110797. DOI: 10.1016/j.mcat.2020.110797

[75] Wang F, Xing Y, Su W, Fan Y, Zhang C, Yang L. Nanoarchitectonics of hierarchical FeCdS-CdS nanorods and Ni-Salens for high-efficiency photocatalytic hydrogen production.

Journal of Alloys and Compounds. 2022; **926**:166907. DOI: 10.1016/j.jallcom.2022.166907

[76] Zubair M, Marie E, Vanhaecke M, Svenum I, Magnus R, Yang J. Core-shell particles of C-doped CdS and graphene: A noble metal-free approach for efficient photocatalytic H_2 generation. Green Energy & Environment. 2020;**5**:461-472. DOI: 10.1016/j.gee.2020.10.017

[77] Zou Y, Guo C, Cao X, Zhang L, Chen T, Guo C, et al. Synthesis of CdS/CoP hollow nanocages with improved photocatalytic water splitting performance for hydrogen evolution. Journal of Environmental Chemical Engineering. 2021;**9**:106270. DOI: 10.1016/j.jece.2021.106270

[78] Yuan C, Lv H, Zhang Y, Fei Q, Xiao D, Yin H, et al. Three-dimensional nanoporous heterojunction of CdS/np-rGO for highly efficient photocatalytic hydrogen evolution under visible light. Carbone New York. 2023;**206**:237-245. DOI: 10.1016/j.carbon.2023.02.022

[79] Bashiri N, Brösigke G, Gioria E, Schmidt J, Konrad M, Oliveira RL, et al. Core–Shell catalyst particles for tandem catalysis: An experimental/numerical approach towards optimal design. Chemical Engineering Journal. 2024;**495**: 153080. DOI: 10.1016/j.cej.2024.153080

[80] Ding M, Ao W, Xu H, Chen W, Tao L, Shen Z, et al. Facile construction of dual heterojunction CoO@TiO_2/MXene hybrid with efficient and stable catalytic activity for phenol degradation with peroxymonosulfate under visible light irradiation. Journal of Hazardous Materials. 2021;**420**:126686. DOI: 10.1016/j.jhazmat.2021.126686

[81] Wang W, He J, Deng J, Chen X, Yu C. Electro-, thermo-, and photocatalysis of versatile nanocomposites toward tandem process. IScience. 2024;**27**:108781. DOI: 10.1016/j.isci.2024.108781

[82] Liu M, Wu X, Dyson PJ. Tandem catalysis enables chlorine-containing waste as chlorination reagents. Nature Chemistry. 2024;**16**:700-708. DOI: 10.1038/s41557-024-01462-8

[83] Aslan Y, Buyuksarac A, Erik NY, Aydemir A, Ates A. Geophysical investigation and hydrocarbon potential of Cankiri-Corum Basin, Turkey. Journal of Petroleum Science and Engineering. 2013;**110**:94-108. DOI: 10.1016/j.petrol.2013.09.004

Chapter 6

Recent Advancements on Key Electrode Materials of Seawater Electrolysis for Hydrogen Production

Min Wei and Xia Sheng

Abstract

Direct seawater electrolysis offers a promising route for sustainable hydrogen production by utilizing abundant seawater and renewable energy. However, the complex chemical composition of seawater, particularly high chloride ion concentrations, poses severe challenges during direct electrolysis, including anode corrosion and competitive side reactions like the chlorine evolution reaction (CER), which impair efficiency and durability. Consequently, stringent demands are focused on developing advanced electrocatalysts resistant to corrosion, selectively promoting the oxygen evolution reaction (OER) over CER, and designing a novel overall system. This chapter reviews the critical breakthroughs and technological directions in seawater electrolysis materials, incorporating recent research advancements from four key perspectives. We outline the fundamental principles of seawater electrolysis and then address the critical challenges of seawater electrolysis: competitive anodic reactions (especially CER) and halogen-induced corrosion. Subsequently, we detail the catalyst design strategies, including (1) constructing Cl^--blocking protective layers, (2) enhancing catalytic activity and selectivity, and (3) coupling non-chlorine oxidation reactions. Future research should focus on designing corrosion-resistant, high-activity electrodes and integrated systems while unraveling multi-reaction mechanisms and standardizing tests in real seawater.

Keywords: seawater electrolysis, hydrogen production, OER, CER, corrosion, strategies for materials engineering

1. Introduction

Hydrogen, a clean and efficient energy carrier producing only water upon combustion without greenhouse gas emissions, is widely regarded as a core component of future energy systems, poised to replace fossil fuels in transportation and industry. Among hydrogen production methods, water electrolysis is key for generating green hydrogen, facilitating the global transition toward green energy [1]. However, industrial-scale electrolysis demands substantial high-purity freshwater, posing critical challenges in water-scarce regions. Furthermore, complete water decomposition necessitates complex, costly purification systems. Consequently, given global freshwater scarcity and rising hydrogen demand, direct seawater electrolysis powered by renewable energy sources has gained significant research attention for sustainable hydrogen production [2].

Currently, seawater electrolysis uses two methods: indirect and direct. The indirect method is more mature than the direct one, but it desalinates and purifies seawater first, adding complexity and cost and reducing its large-scale viability. In contrast, direct seawater electrolysis uses raw seawater without pretreatment, potentially lowering energy use and infrastructure needs. The inherent electrolytes in seawater (e.g., Na^+, Cl^-) also enhance ionic conductivity, potentially boosting efficiency [3]. Despite these advantages, direct seawater electrolysis faces major hurdles. First, high chloride levels cause severe corrosion of electrodes and components, shortening lifespan. Second, chloride ions compete at the anode, generating hypochlorite *via* the CER instead of OER, lowering Faradaic efficiency and producing corrosive byproducts. Third, cations like Mg^{2+} and Ca^{2+} precipitate on the cathode, blocking active sites and hindering the hydrogen evolution reaction (HER) [4].

Addressing these limitations requires innovative materials and system designs. Research focuses on developing corrosion-resistant electrodes and advanced electrocatalysts. Tailoring catalyst composition, morphology, and interfaces aims to improve OER

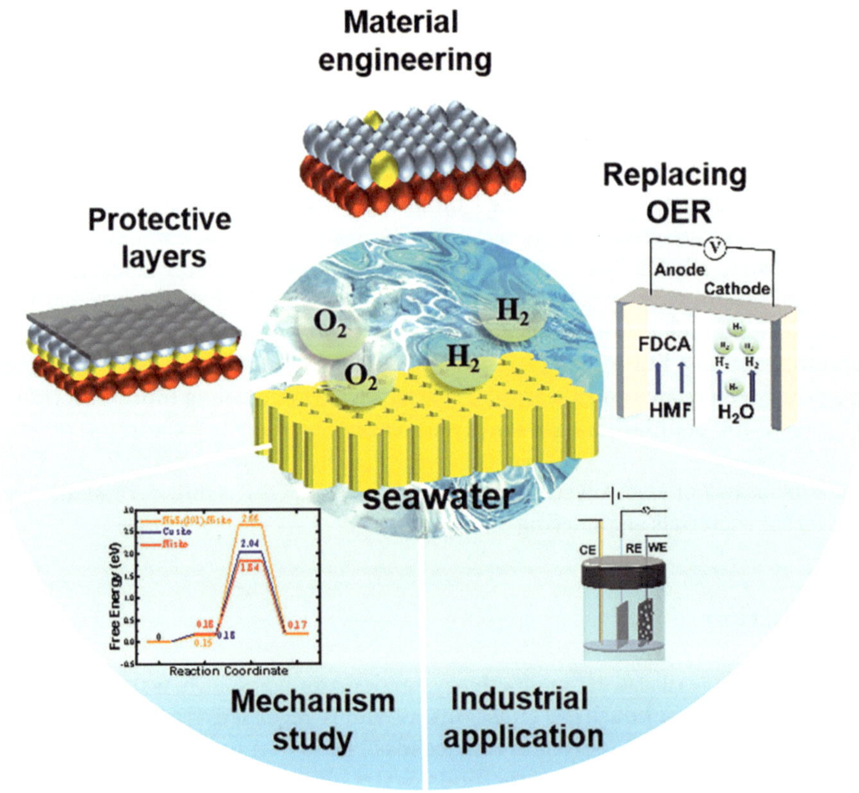

Figure 1.
Brief description of catalyst design as well as advancements and tendencies in direct seawater electrolysis.

selectivity over CER, reduce corrosion, and minimize precipitate deposition [2]. Novel system designs, like membrane-free electrolyzers and selective ion-conducting separators, are also being explored to enhance durability and efficiency [5]. Overcoming these barriers could make direct seawater electrolysis a scalable route for green hydrogen, especially for coastal and offshore renewable energy. Herein, this chapter focuses on recent advancements in key electrode materials for seawater electrolysis. We outline the fundamental principles of seawater electrolysis and then address critical challenges of seawater electrolysis. Subsequent sections detail the catalyst design strategies and prospect the future development trends of electrode materials (see **Figure 1**).

2. Mechanism of seawater electrolysis for hydrogen production

Seawater electrolysis follows a mechanism analogous to freshwater electrolysis, involving the oxygen evolution reaction (OER) at the anode and the hydrogen evolution reaction (HER) at the cathode. The specific reaction pathway depends on pH: acidic electrolytes proceed *via* Eq. (1) and Eq. (2), while neutral or alkaline conditions follow Eq. (3) and Eq. (4). The overall reaction is given by Eq. (5).

$$2H_2O \rightarrow O_2\uparrow + 4H^+ + 4e^- \left(E_0 = 1.23\,V\ vs.SHE, pH = 0\right) \tag{1}$$

$$4H^+ + 4e^- \rightarrow 2H_2 \left(E_0 = 0\,V\ vs.SHE, pH = 0\right) \tag{2}$$

$$4OH^- \rightarrow O_2\uparrow + 2H_2O + 4e^- \left(E_0 = 0.41\,V\ vs.SHE, pH = 14\right) \tag{3}$$

$$4H_2O + 4e^- \rightarrow 2H_2 + 4OH^- \left(E_0 = -0.828\,V\ vs.SHE, pH = 14\right) \tag{4}$$

$$\text{The overall reaction equation}: 2H_2O(l) \rightarrow 2H_2\uparrow + O_2\uparrow \tag{5}$$

Since natural seawater is alkaline (pH $\approx$ 8), its electrolysis therefore follows the neutral/alkaline pathway mechanism, as shown in Eq. (3) and Eq. (4). In this case, OH^- ions are the primary reactant for OER under alkaline conditions. The corresponding OER and HER mechanisms are outlined below.

2.1 The mechanism of OER

The oxygen evolution reaction (OER) at the anode involves a complex four-electron transfer process [1, 2]. Currently, researchers propose various mechanisms for water oxidation, notably the adsorbate evolution mechanism (AEM) and lattice-oxygen-mediated mechanism (LOM) (**Figure 2a-b**). Specifically, in alkaline media (Eq. (6)-Eq. (9)), H_2O adsorption occurs on the catalyst's active site (*), followed by a proton-coupled electron transfer step, and then direct coupling with another OH^-/ H_2O, and finally O_2 release.

$$OH^- + {}^* \rightarrow OH^* + e^- \tag{6}$$

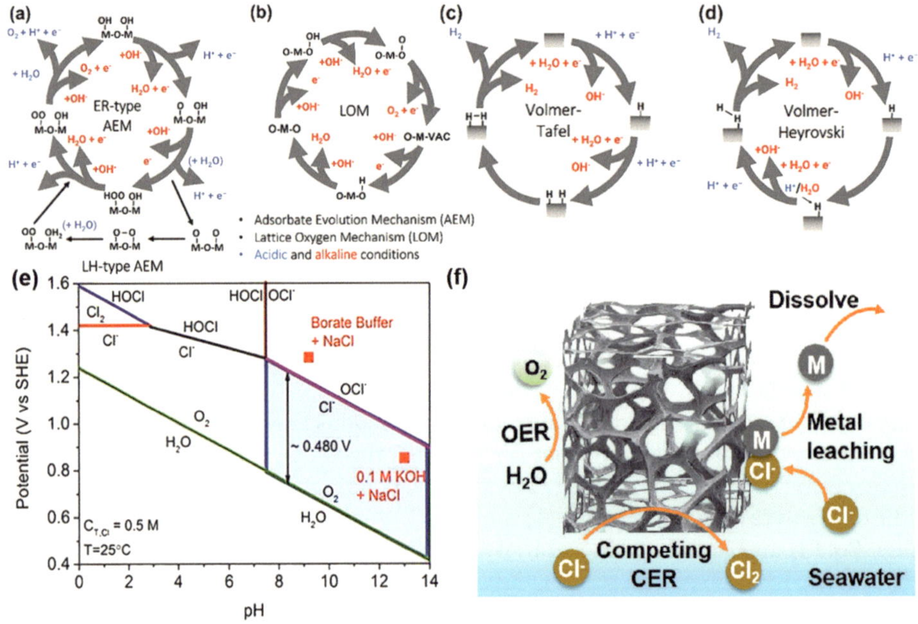

Figure 2.
Reaction mechanisms OER (a-b) and HER (c-d) in the acidic (blue) or alkaline (red) medium [1]. Copyright 2022 Royal Society of Chemistry. (e) Pourbaix diagram including H_2O/O_2 and $Cl^-/Cl_2/HOCl/ClO^-$ redox pairs [5]. Copyright 2023 American Chemical Society. (f) Cl^- corrosion on metal sites in seawater.

$$OH^* + OH^- \rightarrow O^* + e^- + H_2O \quad (7)$$

$$O^* + OH^- \rightarrow HOO^* + e^- \quad (8)$$

$$HOO^* \rightarrow {}^* + O_2(g) + e^- + H^+ \quad (9)$$

The OER pathway involves adsorption/desorption of oxygen-containing intermediates (notably OH*, O*, and OOH*) on the catalyst surface and exhibits slow kinetics. Consequently, the binding strength between the catalyst and these intermediates is a key descriptor for evaluating OER activity.

2.2 The mechanism of HER

HER at the cathode during seawater electrolysis is a two-electron transfer process (**Figure 2c-d**) [1]. Typically, in neutral/alkaline media (Eq. (10)-Eq. (12)), HER proceeds *via* three steps: First, the step that mainly occurs is the Volmer step (an electrochemical step), where H_2O gains electrons on the catalyst surface to form hydrogen adsorption intermediates H* and OH^-. Subsequently, H* combines with another water molecule and an electron to produce H_2 (Heyrovsky step, electrochemical desorption step), or two activated H^* combine to produce H_2 (Tafel step, composite desorption step), or both occur simultaneously [1].

$$\text{Volmer step}: H_2O + e^- + ^* \rightarrow H^* + OH^- \quad (10)$$

$$\text{Heyrovsky step}: H^* + H_2O + e^- \rightarrow H_2 + OH^- + ^* \quad (11)$$

$$\text{Tafel step}: 2H^* \rightarrow H_2 \uparrow \quad (12)$$

These elementary steps reveal that HER requires the H^* intermediate (supplied by H_2O dissociation, not H_3O^+), consequently imposing a high thermodynamic barrier due to the additional energy for water splitting. Notably, H^* formation is more challenging in alkaline media than acidic pathways. According to the Sabatier principle [3], optimal HER activity occurs when the hydrogen adsorption free energy (ΔG_{H^*}) approaches zero. Thus, ΔG_{H^*} is a key descriptor of catalyst activity for HER in seawater electrolysis.

3. Challenges of seawater electrolysis for hydrogen production

Unlike pure water, the complex composition of natural seawater triggers competitive reactions, catalyst corrosion, and active site blockage during electrolysis [6]. Furthermore, its near-neutral pH—unlike common acidic/alkaline electrolytes—results in slowing reaction kinetics. Therefore, it is necessary to clarify the mechanism of the challenges that harm the electrode materials.

3.1 Anodic competitive reaction

Natural seawater contains abundant Cl^-. Critically, CER involves only a two-electron transfer with one catalytic intermediate, making it thermodynamically favorable compared to the complex four-electron OER. Thus, CER competes with OER at the anode during seawater electrolysis, significantly reducing OER selectivity.

CER is influenced by pH, applied potential, and Cl^- concentration, generating toxic and corrosive products (e.g., Cl_2, HClO, ClO^-). Moreover, the Pourbaix diagram (**Figure 2e**) shows reduced potential difference between the CER and OER as pH decreases [5]. Specifically, below pH 3.0, Cl_2 dominates *via* CER (Eq. 13); at pH 3.0–7.5, hypochlorous acid (HClO) forms; while above pH 7.5, hypochlorite (ClO^-) becomes primary (Eq. 14), often termed the hypochlorite formation reaction.

$$2Cl^- \rightarrow Cl_2 + 2e^-,\ E_0 = +1.36\ \text{V vs.SHE, pH} = 0 \quad (13)$$

$$Cl^- + 2OH^- \rightarrow ClO^- + H_2O + 2e^-,\ E_0 = +0.89\ \text{V vs.SHE, pH} = 14 \quad (14)$$

Here, E_0 represents the standard electrode potential, and SHE indicates the standard hydrogen electrode.

Notably, at pH ≥ 7.5, the potential difference between CER and OER reaches its maximum value (480 mV), indicating maximal OER selectivity under alkaline conditions. Consequently, seawater electrolysis typically operates in alkaline/weakly alkaline media. However, adding alkali (e.g., KOH) raises operational costs, while anodic OER consumes OH^-, inducing localized acidic zones. This pH heterogeneity accelerates catalyst degradation and reduces CER's energy barrier. Therefore, developing highly selective OER catalysts is essential to suppress corrosive byproduct (e.g., ClO^-/ClO_3^-) generation during seawater electrolysis.

3.2 Chlorine corrosion

The high Cl^- concentration in seawater and corrosive CER byproducts (Cl_2, HClO) severely corrode catalysts, as illustrated in **Figure 2f** [4]. Moreover, under anodic potential, Cl^- adsorbs onto metal active sites, forming intermediates (Eq. 15). These intermediates subsequently coordinate with additional Cl^- to generate soluble MCl_x^- complexes (Eq. 16), which dissolve from the electrode. Ultimately, dissolved MCl_x^- complexes hydrolyze into inactive hydroxide precipitates (Eq. 17), causing catalyst deactivation. Therefore, designing corrosion-resistant anode catalysts is imperative for seawater electrolysis.

$$M + Cl^- \rightarrow MCl_{ads} + e^- \tag{15}$$

$$MCl_{ads} + Cl^- \rightarrow MCl_X^- \tag{16}$$

$$MCl_X^- + OH^- \rightarrow M(OH)_X + C \tag{17}$$

Additionally, the cations in seawater (e.g., Mg^{2+}, Ca^{2+}, K^+, Na^+,) interact with the lattice oxygen of the catalyst during the reaction, potentially replacing host cations and causing catalyst poisoning. Furthermore, HER-induced cathode alkalinity triggers hydroxide scaling ($Mg(OH)_2/Ca(OH)_2$), blocking the active sites. Subsequently, corroded electrodes develop pitting, cracking, and delamination, which reduce the catalytic efficiency and active area of the electrode. This damage increases the local current density, accelerating corrosion [7]. Consequently, cation-induced degradation severely compromises electrode activity and stability in seawater electrolysis.

4. Key strategies for the design of electrocatalysts

It is precisely because of the various challenges posed by seawater that designing effective electrocatalysts for both cathode and anode imposes more demanding criteria than freshwater systems. The performances of catalysts for direct seawater electrolysis are critically governed by active site density, intrinsic conductivity, and reaction kinetics [4]. Drawing on the current research advancements, this section outlines effective structural design strategies of electrocatalysts to specifically enhance chloride corrosion resistance and improve long-term durability for seawater electrolysis.

4.1 Establishing protective layers

In seawater electrolysis, Cl^- ions can corrode both the catalyst and substrate. While maintaining catalyst activity, rationally designing a surface protective layer represents an effective strategy to enhance the OER selectivity and extend catalyst lifespan by shielding active sites/substrates from Cl^- corrosion. Common approaches include applying corrosion-resistant coatings, establishing negative charge repulsion barriers, and modulating the local microenvironment, as shown in **Figure 3a** [5].

4.1.1 Coating with corrosion-resistant layers

Since Bennett's seminal 1980 discovery that MnO_x exhibits a high OER selectivity of 99% in acidic brine electrolysis [11], it has been widely employed as a Cl^--blocking protective layer. For instance, Hashimoto et al. developed IrO_x/Ti-supported MnO_x electrodes, achieving near-100% OER selectivity and exceptional stability across diverse

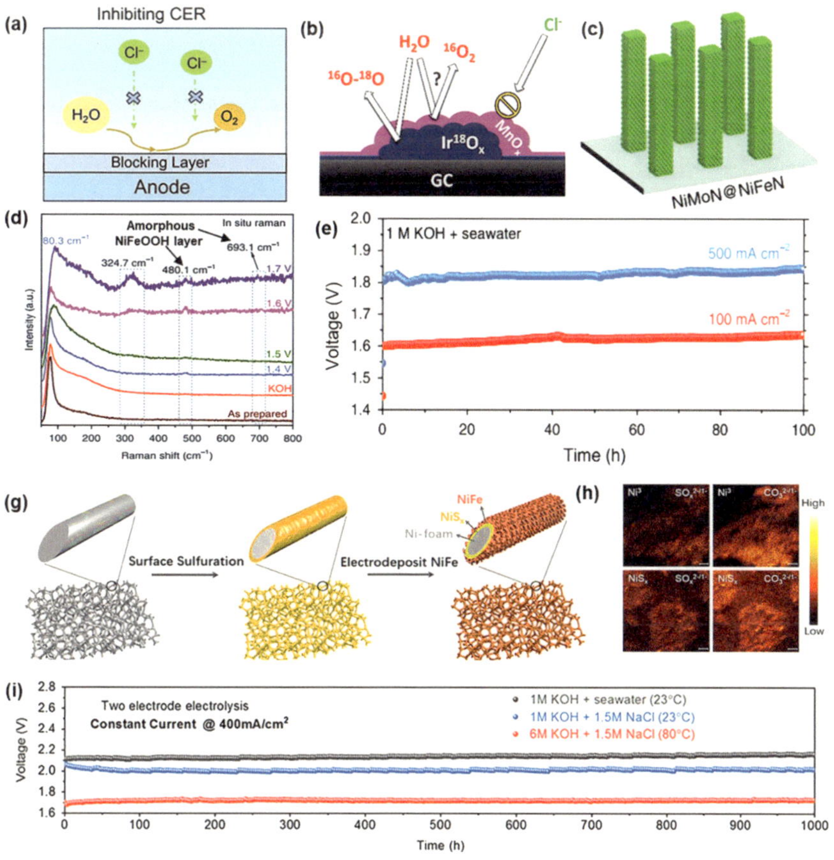

Figure 3.
(a) Constructing a blocking layer against Cl^- and CER [5]. Copyright 2023 American Chemical Society. (b) Deposition Structure of MnOx Deposition on MnO_x/IrO_x/GC [8]. Copyright 2018 American Chemical Society. (c) Schematic illustration, (d) in situ *Raman spectra at various potentials for the OER process, and (e) stability test in seawater of the 3D NiMoN@NiFeN catalyst [9]. Copyright 2019 Springer Nature (http://creativecommons.org/licenses/by/4.0/). (g) Construction and structure, (h) TOF-SIMS mapping of $SO_x^{2-/1-}$ and $CO_3^{2-/1-}$ fragments and (i) durability tests of the NiFe/NiSx-Ni foam anode for seawater splitting [10]. Copyright 2019 PNAS (https://creativecommons.org/licenses/by-nc-nd/4.0/).*

seawater electrolysis conditions [12]. Similarly, Johannes et al. electrodeposited ultrathin MnO_x films (~5–20 nm) onto hydrous IrO_x/glass carbon in acidic chloride solution (pH ~ 0.9) [8]. This coating reduced CER selectivity from 86% to less than 7% in 30 mM Cl^-, indicating high OER selectivity. Crucially, the catalytically inert MnO_x layer functions as a selective diffusion barrier: it blocks Cl^- access to underlying IrO_x active sites while permitting transport of H_2O, H^+, and O_2 essential for OER activity (**Figure 3b**).

Beyond manganese oxides, corrosion-resistant hydroxides, nitrides, and sulfides leverage unique electronic structures to *in situ* generate protective layers during operation. Ren et al. engineered an Ni foam-supported 3D core-shell NiMoN@NiFeN electrocatalyst for alkaline seawater electrolysis (**Figure 3c**) [9]. Here, conductive NiMoN nanorods provide a large surface area and efficient charge transfer, while outer NiFeN nanoparticles evolve *in situ* into amorphous NiFe (oxy)hydroxide layers during OER (**Figure 3d**). These layers not only enhance OER selectivity but also confer chloride corrosion resistance. Concurrently, the hierarchical porosity facilitates seawater diffusion and H_2/O_2 bubble release, collectively boosting performance and stability (**Figure 3e**).

4.1.2 Constructing Cl-repelling layer on the surface of the catalyst

Constructing a negatively charged protective layer on the surface of electrocatalysts effectively repels Cl^- *via* electrostatic repulsion, mitigating corrosion during seawater electrolysis while enhancing OER selectivity.

Multiple studies confirm that polyatomic sulfate ions repel Cl^- *via* electrostatic interactions, consequently enhancing the corrosion resistance of the catalyst during seawater electrolysis. For instance, Kuang et al. designed a NiFe hydroxide/NiS_x-Ni foam multilayer anode for solar-driven alkaline seawater electrolysis (**Figure 3g**) [10]. *In situ* formed sulfate/carbonate-rich passivation layers during operation provided superior Cl^- rejection and corrosion resistance (**Figure 3h**). NiFe/NiSx-Ni only required 510 mV overpotential to achieve 400 mA/cm^2 with no activity decay after 1000 h (**Figure 3i**). Similarly, Gao et al. created a karst-like Ni foam-supported electrode with metallic Ni valleys and Ni/α-$Ni(OH)_2$ heterostructure "towers" [13]. Slightly negatively charged Ni^0 atoms facilitated Ni-H_{ad} adsorption, while positively charged $Ni(OH)_2$ promoted hydroxide acceptance and water dissociation, accelerating OER kinetics. Critically, XPS/Raman analyses revealed the *in situ* formed sulfate passivation layers on the electrode that blocked Cl^- ingress, preventing corrosion in seawater. Bahuguna et al. synthesized a catenated sulfur-type Ni polysulfide electrocatalyst ($NiS_2pS_x^{surface}$) for neutral seawater splitting [14]. At 10 mA/cm^2, its overpotential was 320 mV lower than IrO_2 and below the chlorine evolution threshold. This ion selectivity originated from steric/electrostatic shielding of Ni active sites by large polysulfide chains (pS_n^{2-}/S_2^- = 2.1), with *in situ* surface evolution further enhancing the intrinsic chlorine-phobicity. Furthermore, Fan et al. developed an earth-abundant CoFe-layered double hydroxide (LDH) electrocatalyst with surface-anchored graphene quantum dots (GQDs) [6]. Density functional theory (DFT) calculations confirmed that the ΔG values of the rate-determining step for CER significantly exceed those of OER at Co sites, explaining the observed approximately 100% OER selectivity on CoFe-Ci@GQDs/NF. Critically, introducing CO_3^{2-} into LDH interlayers established Cl^--repelling interlayer galleries, while surface GQDs enabled electrostatic Cl^- repulsion from active sites. This dual-scale modification enhanced corrosion resistance, achieving more than 2800 h stability in alkaline seawater at 1.25 A/cm^2.

4.1.3 Regulating the composition of the electrolyte

Incorporating electrolyte additives (e.g., SO_4^{2-}, PO_4^{3-}) mitigates chloride corrosion by preferentially enriching anions on the OER catalyst surface. This generates electrostatic repulsion against Cl^-, complementing surface barrier layers that similarly repel chloride ions.

Chen et al. demonstrated that sulfate (SO_4^{2-}), as a cost-effective electrolyte additive, significantly enhances the stability of Ni anode in alkaline seawater electrolysis [15]. Theoretical and *in situ* analyses confirm SO_4^{2-} preferentially adsorbs on anodes surface, forming a negatively charged layer that electrostatically repels Cl^-. Notably, using NiFe-LDH/NF anodes in SO_4^{2-}-supplemented electrolyte enabled stable hydrogen production at 400 mA/cm^2 for 1000 h (simulated seawater) and 500 h (real seawater)—fivefold longer than sulfate-free systems.

Similarly, Yu et al. reported phosphate ions (PO_4^{3-}) as electrolyte additives for stable alkaline seawater splitting [16]. *In situ* studies and theoretical simulations reveal PO_4^{3-} provides dynamic corrosion protection *via* (1) selective Cl^- repulsion without impeding OH^- diffusion, (2) suppression of transition metal dissolution, and (3) local pH buffering through PO_4^{3-}/HPO_4^{2-} transformation, compensating for rapid OH^- consumption at high currents. Consequently, phosphate-assisted systems achieved over 500 h stability at 0.5 A/cm^2 in real seawater.

4.2 Enhancing catalytic activity and selectivity

The Pourbaix diagram (**Figure 1e**) shows a 480 mV potential window between the OER and CER in seawater electrolysis, confirming the thermodynamic preference of OER [2]. Therefore, designing high-performance OER electrocatalysts capable of achieving industrial current densities below 1.71 V can effectively suppress CER, ensuring selective oxygen evolution. To address this, enhancing both intrinsic activity and catalytic selectivity is essential. Key strategies include structural engineering to maximize active site exposure, defect engineering to introduce vacancies, heteroatom doping to tailor electronic properties, and interface engineering to optimize charge transfer. This section systematically examines these approaches, highlighting their synergistic roles in improving electrocatalytic seawater splitting for efficient hydrogen production.

4.2.1 Structural engineering

Rational design of catalyst structure and morphology—including particle size reduction, enhanced mass transport channels, and improved conductivity—effectively boosts catalytic activity. For instance, Li et al. engineered a multilayered OER electrode comprising amorphous FeOOH-coated Mo-doped $Co_{0.85}Se$ nanosheets on 3D macroporous Ni foam, forming a crystalline-amorphous heterostructure (**Figure 4a**) [17]. This configuration not only increased active sites through electronic modulation but also enhanced mass transfer *via* superhydrophilic/superaerophobic surfaces. Crucially, the architecture *in situ* transformed into a chloride-resistant metal (oxy)hydroxide layer enriched with polyanions (MoO_x^{n-}/SeO_x^{n-}) (**Figure 4b**). Consequently, NF/$(CoMo)_{0.85}Se$@FeOOH achieved low overpotentials of 266, 287, and 311 mV at 50, 100, and 200 mA/cm^2 in alkaline simulated seawater, maintaining stability for 140 h at 50 mA/cm^2.

Meanwhile, dispersing catalysts on carbon materials mitigates nanoparticle aggregation while enriching active sites and promoting seawater electrolysis. Wang

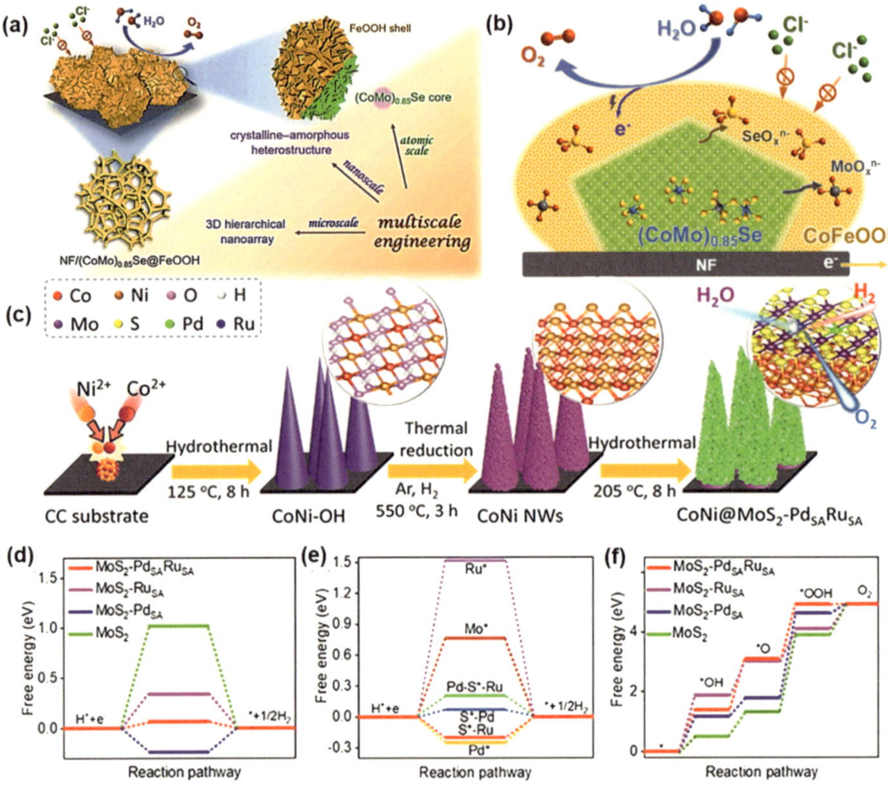

Figure 4.

Strategies of enhancing catalytic activity and selectivity. (a) Synthetic route and (b) the mechanism of resistance to chloride ion corrosion of NF/(CoMo)$_{0.85}$Se@FeOOH [17]. Copyright 2023 American Chemical Society. (c) The illustration of CoNi@MoS$_2$-Pd$_{SA}$Ru$_{SA}$ hybrid, ΔG_{H^*} values of (d) different materials and (e) different sites for HER, and (f) OER pathway for CoNi@MoS$_2$-Pd$_{SA}$Ru$_{SA}$ [18]. Copyright 2023 American Chemical Society.

et al. synthesized Ni nanoparticles/N-doped carbon sheets *via* pyrolysis of Ni-based triazole-benzoic acid complexes, with HRTEM confirming the uniform distribution of Ni nanoparticles with 4.8–8.8 nm Ni nanoparticles encapsulated in N-doped carbon sheets [19]. The synergism of Ni/N-doped carbon enhanced acidic/alkaline OER and HER, enabling an electrolyzer constructed by Ni/N-C/NiFe foam requiring only 1.56 V (freshwater) and 1.62 V (seawater) at 10 mA/cm^2.

Furthermore, two-dimensional metal–organic frameworks (2D MOFs) exhibit exceptional OER activity due to their planar conjugated layers, uniform pores, large surface area (over 1000 m^2/g), and high conductivity (exceeding 10^3 S/cm), which collectively enhance mass transport and reaction kinetics [20].

4.2.2 Defect engineering

Defect engineering effectively optimizes catalysts' intrinsic activity by regulating the concentration and properties of defects, which modulates electronic structures to lower reaction energy barriers and exposes coordination-unsaturated active sites. Specifically, Liu et al. synthesized needle-like Fe-doped Ni$_2$P with P vacancies (Fe-Ni$_2$Pv), demonstrating superior bifunctional HER/OER activity in alkaline seawater electrolysis [21]. This catalyst achieved remarkably low voltages of 1.68 and

1.73 V at 1 and 3 A/cm^2, respectively. Theoretical and experimental analyses confirmed that Fe doping and P vacancies accelerate OER-active species reconstruction while optimizing hydrogen adsorption free energy (ΔG_{H^*}) for HER, with vacancies enhancing electrical conductivity greatly.

Moreover, oxygen vacancies critically modify the electronic structures and surface properties of electrocatalysts. Meng et al. engineered MOF-derived $Pt_2/Ni(OH)_2/NF$ nanosheets-nanoparticles featuring abundant O vacancies (Ov), strong interfacial metal-support interaction (SMSI), and superhydrophilicity [22]. EPR spectra verified Ov presence, which weakened the adsorption of H* and facilitated efficient H_2O adsorption. Theoretical studies revealed Ov optimizes hydrogen spillover from $Ni(OH)_2$ to Pt, yielding exceptional HER activity.

4.2.3 Doping engineering

Doping engineering introduces heteroatoms into catalyst matrices to modulate electronic structures and optimize the desorption free energy of key oxygen intermediates (e.g., O*, OOH*) during the electrocatalytic reaction. Non-noble metal dopants (Fe [23], Mo [24], V [25], and Cr [26]) are particularly promising due to their efficiency, stability, and cost-effectiveness. For example, Long et al. synthesized Fe-doped phosphide (Fe-Ni_5P_4) from NiFc-MOF precursors [23]. Theoretical calculations revealed Fe doping simultaneously reduces the OER energy barriers ($O^* \rightarrow OOH^* \rightarrow O_2^*$) and optimizes the intermediate adsorption of HER (H_2O^*/H^*). This catalyst achieved remarkably low overpotentials of 230 mV (OER) and 250 mV (HER) at 500 mA/cm^2. The Fe-Ni_5P_4||Fe-Ni_5P_4 electrolyzer sustained robust alkaline seawater splitting at 1 A/cm^2 for 1170 h. Furthermore, trace noble metals (Au [27], Ru [25], Pd [18], Ir [28]) enhance the catalytic activity through electronic modulation. Islam et al. developed Pd and Ru dual-doped MoS_2-shelled CoNi alloy nanowires (CoNi@MoS_2-$Pd_{SA}Ru_{SA}$) (**Figure 4c**) [18]. Theoretical studies confirmed dual-atomically dispersed Pd and Ru doping restructured electronic environments and induced synergistic effects while accelerating HER (**Figure 4d-e**) and OER (**Figure 4f**) kinetics. The resulting electrocatalyst required only 1.54 V to attain 10 mA/cm^2 during overall natural seawater splitting.

4.2.4 Interface engineering

Constructing heterogeneous interfaces enhances catalytic activity through synergistic effects, improving electronic properties, active site accessibility, and interfacial charge transfer. Wei et al. developed 3D/2D MoS_2@W-G heterostructures by fabricating MoS_2 nanoflowers on W oxide-anchored graphene (W-G) nanosheets, demonstrating exceptional HER performance [29]. The catalyst achieved Faradaic efficiencies of 96% (acidic) and 94% (alkaline) in real seawater, with ultralow overpotentials of 78 and 94 mV at 10 mA/cm^2. XPS/Raman confirmed the formation of C···W-O···S-Mo interfaces, while DFT calculations revealed these interfaces synergistically optimized hydrogen adsorption/desorption. This accelerated the Volmer-Heyrovsky pathway, significantly boosting HER kinetics.

4.3 Coupling chlorine-free oxidation reaction

Overall, water electrolysis suffers from slow kinetics involving multistep proton-coupled electron transfers and a high thermodynamic barrier (1.23 V vs. RHE).

Substituting OER with thermodynamically favorable electro-oxidation reactions avoids competitive CER in seawater. This strategy reduces energy consumption while improving hydrogen production efficiency and enabling energy-efficient electrolysis.

4.3.1 Biomass molecules oxidation

Biomass small-molecule oxidation often exhibits high selectivity at low current densities. Replacing OER with such reactions avoids competitive CER and obtains high value-added products under industrially viable conditions. Recent studies demonstrate seawater electrolysis coupled with oxidation of ethylene glycol [30], glycerol [31], methanol [32], ethanol [33], 5-hydroxymethylfurfural [34], and acrylamide [35].

Our work developed an Fe/NiB/CF catalyst for 5-hydroxyfurfural oxidation (HMFOR), achieving 100 mA/cm^2 at 1.46 V in seawater while converting HMF to 2,5-furandicarboxylic acid with ~100% yield (**Figure 5a**) [36]. By coupling HMFOR with HER using Fe-electrodeposited carbonized wood (Fe-CW), the Fe-CW||Fe/NiB/CF cell attained 100 mA/cm^2 at 1.47 V. This strategy enables energy-efficient coproduction of hydrogen and high-value chemicals, advancing industrial biomass-coupled electrolysis.

4.3.2 Inorganic molecules oxidation

By coupling such oxidation reactions with the hydrogen evolution reaction (HER) in seawater electrolysis, this strategy simultaneously addresses water pollution and enables energy-efficient hydrogen production.

Common pollutants in industrial waste gas and wastewater, such as hydrogen sulfide or sulfion, hydrazine, and urea, are toxic substances that can lead to environmental pollution and pose significant risks to human health. Electro-oxidation represents a viable treatment approach capable of rapidly degrading these toxic substances without the need for additional oxidants or intricate separation techniques. Integrating the oxidation of these inorganic compounds with HER for seawater electrolysis will

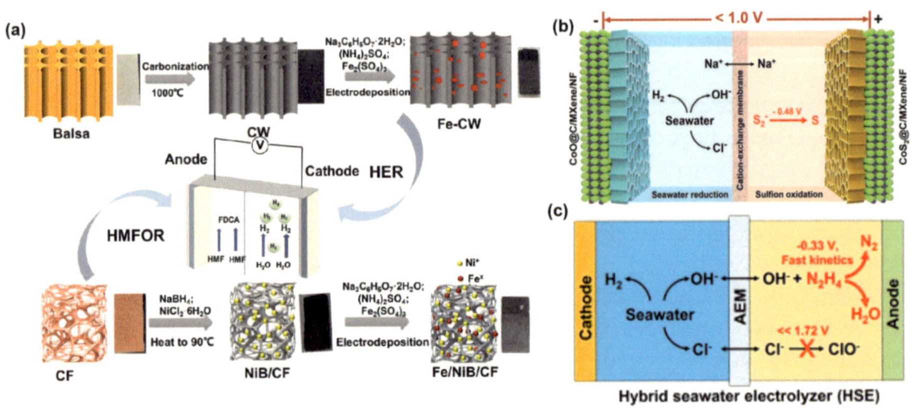

Figure 5.
The construction of a hybrid seawater electrolyzer by coupling seawater reduction with the oxidation reaction of (a) HMF [36] (Copyright 2025 the authors), (b) sulfion [37] (Copyright 2022 Wiley-VCH), and (c) hydrazine [38] (Copyright 2021 Springer Nature: https://creativecommons.org/licenses/by/4.0/).

not only address the challenge of water pollution but also enable hydrogen production with economic and environmental advantages (**Figure 5b-c**). Transition metal catalysts, including Co [37], Ni [38], Mo [39], and Fe [40], drive these processes, offering dual environmental and economic benefits.

5. Conclusions and perspectives

This chapter summarizes the recent advancements in key electrode materials for seawater electrolysis. We first outline the fundamental principles of seawater electrolysis and then address critical challenges of seawater electrolysis: competitive anodic reactions (especially CER) and halogen-induced corrosion. Subsequent sections detail the catalyst design strategies, including (1) constructing Cl^--blocking protective layers, (2) enhancing catalytic activity and selectivity, and (3) coupling non-chlorine oxidation reactions. Despite recent progress in electrode materials with high selectivity and robust stability, direct seawater electrolysis remains nascent and full of challenges:

1. Balance of corrosion resistance and activity. Current protective layers (e.g., polyanion-rich surfaces) often sacrifice conductivity or active site density. Developing advanced protective coatings, engineering multiscale-structured catalysts, and integrating defects/dopants are effective strategies.

2. Reduction of lab-industry gaps. IR-corrected and simulated seawater often overestimate catalyst activity in the laboratory, causing significant performance gaps in actual industrial applications. Bridging the lab-industry performance gap requires establishing standardized testing protocols using natural seawater to accelerate the practical deployment of efficient and durable seawater electrolysis technologies.

3. Exploration of multi-reaction coupling mechanisms. Fundamental understanding of interfacial processes in hybrid systems (e.g., biomass oxidation coupling with HER) is lacking. Combining *in situ* techniques such as X-ray absorption fine structure (XAFS), Raman, and XRD with theoretical calculations to establish the competing ion adsorption and charge transfer kinetics.

Acknowledgements

This research was financially provided by the Top-Notch Personnel Fund of Henan Agricultural University (30501033), the Henan Provincial Joint Fund (222103810023), the Fund of Longzihu New Energy Laboratory (LZGLH2023011), and the Natural Science Foundation Project of the Henan Province Department of China (252300420779).

Conflict of interest

The authors declare no conflict of interest.

Author details

Min Wei[1] and Xia Sheng[1,2]*

1 College of Sciences, Henan Agricultural University, Zhengzhou, China

2 Longzihu New Energy Laboratory, Zhengzhou, China

*Address all correspondence to: shengxia@henau.edu.cn

References

[1] Chatenet M, Pollet BG, Dekel DR, Dionigi F, Deseure J, Millet P, et al. Water electrolysis: From textbook knowledge to the latest scientific strategies and industrial developments. Chemical Society Reviews. 2022;**51**(11):4583-4762. DOI: 10.1039/D0CS01079K

[2] Liu J, Duan S, Shi H, Wang T, Yang X, Huang Y, et al. Rationally designing efficient electrocatalysts for direct seawater splitting: Challenges, achievements, and promises. Angewandte Chemie International Edition. 2022;**61**(45):e202210753. DOI: 10.1002/anie.202210753

[3] Zhang D, Shi Y, Yin J, Lai J. Recent advances for seawater hydrogen evolution. ChemCatChem. 2024;**16**(14):e202301305. DOI: 10.1002/cctc.202301305

[4] Song L, Chi J, Tang J, Liu X, Xiao Z, Wu Z, et al. Anode design principles for efficient seawater electrolysis and inhibition of chloride oxidation. Chinese Journal of Catalysis. 2024;**66**:53-75. DOI: 10.1016/S1872-2067

[5] He W, Li X, Tang C, Zhou S, Lu X, Li W, et al. Materials design and system innovation for direct and indirect seawater electrolysis. ACS Nano. 2023;**17**(22):22227-22239. DOI: acsnano.3c08450

[6] Fan R, Liu C, Li Z, Huang H, Feng J, Li Z, et al. Ultrastable electrocatalytic seawater splitting at ampere-level current density. Nature Sustainability. 2024;**7**(2):158-167. DOI: 10.1038/s41893-023-01263-w

[7] Sun J, Zhao Z, Li J, Li Z, Meng X. Recent advances in transition metal selenides-based electrocatalysts: Rational design and applications in water splitting. Journal of Alloys and Compounds. 2022;**918**:165719. DOI: 10.1016/j.jallcom.2022.165719

[8] Vos JG, Wezendonk TA, Jeremiasse AW, Koper MTM. MnO_x/IrO_x as selective oxygen evolution electrocatalyst in acidic chloride solution. Journal of the American Chemical Society. 2018;**140**(32):10270-10281. DOI: 10.1021/jacs.8b05382

[9] Yu L, Zhu Q, Song S, McElhenny B, Wang D, Wu C, et al. Non-noble metal-nitride based electrocatalysts for high-performance alkaline seawater electrolysis. Nature Communications. 2019;**10**(1):5106. DOI: 10.1038/s41467-019-13092-7

[10] Kuang Y, Kenney MJ, Meng Y, Hung W-H, Liu Y, Huang JE, et al. Solar-driven, highly sustained splitting of seawater into hydrogen and oxygen fuels. National Academy of Sciences of the United States of America. 2019;**116**(14):6624-6629. DOI: 10.1073/pnas.1900556116

[11] Bennett JE. Electrodes for generation of hydrogen and oxygen from seawater. International Journal of Hydrogen Energy. 1980;**5**(4):401-408. DOI: 10.1016/0360-3199(80)90021-X

[12] Habazaki H, Matsui T, Kawashima A, Asami K, Kumagai N, Hashimoto K. Nanocrystalline manganese-molybdenum-tungsten oxide anodes for oxygen evolution in seawater electrolysis. Scripta Materialia. 2001;**44**(8):1659-1662. DOI: 10.1016/S1359-6462(01)00876-4

[13] Gao X, Chen Y, Sun T, Huang J, Zhang W, Wang Q, et al. Karst

landform-featured monolithic electrode for water electrolysis in neutral media. Energy and Environmental Science. 2020;**13**(1):174-182. DOI: 10.1039/C9EE02380A

[14] Bahuguna G, Filanovsky B, Patolsky F. Pioneering practical direct sea water splitting via an intrinsically-selective chlorine-phobic nickel polysulphide nanostructured electrocatalyst for pure oxygen evolution. Nano Energy. 2023;**111**:108439. DOI: 10.1016/j.nanoen.2023.108439

[15] Ma T, Xu W, Li B, Chen X, Zhao J, Wan S, et al. The critical role of additive sulfate for stable alkaline seawater oxidation on nickel-based electrodes. Angewandte Chemie International Edition. 2021;**60**(42):22740-22744. DOI: 10.1002/anie.202110355

[16] Yu M, Li J, Liu F, Liu J, Xu W, Hu H, et al. Anionic formulation of electrolyte additive toward stable electrocatalytic oxygen evolution in seawater splitting. Journal of Energy Chemistry. 2022;**72**:361-369. DOI: 10.1016/j.jechem.2022.04.004

[17] Li P, Zhao S, Huang Y, Huang Q, Yang Y, Yang H. Multiscale structural engineering of a multilayered nanoarray electrode realizing boosted and sustained oxygen evolution catalysis in seawater electrolysis. ACS Catalysis. 2023;**13**(23):15360-15374. DOI: 10.1021/acscatal.3c03573

[18] Islam M, Nguyen TH, Tran DT, Dinh VA, Kim NH, Lee JH. Bimetallic atom dual-doped MoS2-based heterostructures as a high-efficiency catalyst to boost solar-assisted alkaline seawater electrolysis. ACS Sustainable Chemistry and Engineering. 2023;**11**(17):6688-6697. DOI: 10.1021/acssuschemeng.3c00260

[19] Wang D, Wang X, Qiu H, Tao Y, Yin J, Li J, et al. Synthesis of Ni nanoparticles/N-doped carbon sheets for freshwater and seawater electrolysis. Fuel. 2024;**371**:131995. DOI: 10.1016/j.fuel.2024.131995

[20] Zhong H, Wang M, Chen G, Dong R, Feng X. Two-dimensional conjugated metal–organic frameworks for electrocatalysis: Opportunities and challenges. ACS Nano. 2022;**16**(2):1759-1780. DOI: 10.1021/acsnano.1c10544

[21] Liu X, Yu Q, Qu X, Wang X, Chi J, Wang L. Manipulating electron redistribution in Ni_2P for enhanced alkaline seawater electrolysis. Advanced Materials. 2024;**36**(1):2307395. DOI: 10.1002/adma.202307395

[22] Sun J, Zhang Z, Meng X. Low-Pt supported on MOF-derived $Ni(OH)_2$ with highly-efficiently electrocatalytic seawater splitting at high current density. Applied Catalysis B: Environmental. 2023;**331**:122703. DOI: 10.1016/j.apcatb.2023.122703

[23] Long Y, Jiang P, Liao P, Yang C, Li S, Xian J, et al. Electronic structure regulation by Fe doped Ni-phosphides for long-term overall water splitting at large current density. Small. 2024;**20**(46):2403991. DOI: 10.1002/smll.202403991

[24] Yadav A, Das MR, Deka S. Fine tuning of torus-shaped Mo-doped Ni_2P nanorings for enhanced seawater electrolysis. Small. 2025;**21**(3):2408036. DOI: 10.1002/smll.202408036

[25] Ma Q, Jin H, Xia F, Xu H, Zhu J, Qin R, et al. Ultralow Ru-assisted and vanadium-doped flower-like CoP/Ni_2P heterostructure for efficient water splitting in alkali and seawater. Journal of Materials Chemistry A. 2021;**9**(47):26852-26860. DOI: 10.1039/D1TA08699E

[26] He Y, Yang X, Jiang M, Liu F, Zhang J, Li H, et al. Cr-doped NiFe sulfides nanoplate array: Highly efficient and robust bifunctional electrocatalyst for the overall water splitting and seawater electrolysis. Journal of Colloid and Interface Science. 2025;**680**:1079-1089. DOI: 10.1016/j.jcis.2024.11.072

[27] Luo H, Pan S, Li C, Li P, Ke J, Wang Y, et al. One-dimensional gold-doped monoclinic iridium oxide nanoribbons for high-efficiency acidic oxygen evolution reaction. Catalysis Science and Technology. 2025;**15**(7):2221-2228. DOI: 10.1039/D5CY00014A

[28] Zhai X, Yu Q, Liu G, Bi J, Zhang Y, Chi J, et al. Hierarchical microsphere MOF arrays with ultralow Ir doping for efficient hydrogen evolution coupled with hydrazine oxidation in seawater. Journal of Materials Chemistry A. 2021;**9**(48):27424-27433. DOI: 10.1039/D1TA07328A

[29] Dang VD, Putikam R, Lin M-C, Wei KH. MoS_2 nanoflowers grown on plasma-induced W-anchored graphene for efficient and stable H_2 production through seawater electrolysis. Small. 2024;**20**(2):2305220. DOI: 10.1002/smll.202305220

[30] Mao Q, Deng K, Yu H, Xu Y, Wang Z, Li X, et al. In situ reconstruction of partially hydroxylated porous Rh metallene for ethylene glycol-assisted seawater splitting. Advanced Functional Materials. 2022;**32**(31):2201081. DOI: 10.1002/adfm.202201081

[31] Deng B, Shen J, Lu J, Huang C, Chen Z, Peng F, et al. Ru doping triggering reconstruction of cobalt phosphide for coupling glycerol electrooxidation with seawater electrolysis. Journal of Energy Chemistry. 2025;**100**:317-326. DOI: 10.1016/j.jechem.2024.08.056

[32] Xu K, Liang L, Li T, Bao M, Yu Z, Wang J, et al. $Pt_{1.8}Pd_{0.2}CuGa$ intermetallic nanocatalysts with enhanced methanol oxidation performance for efficient hybrid seawater electrolysis. Advanced Materials. 2024;**36**(31):2403792. DOI: 10.1002/adma.202403792

[33] Jing Y, Lu B, Jiang Y, Zhu K, Ye K. Rational design of V doped CoP/Ni_2P heterostructural catalysts for highly efficient ethanol electrolysis. Journal of Alloys and Compounds. 2025;**1021**:179593. DOI: j.jallcom.2025.179593

[34] Wu J, Zhai Z, Yin S, Wang S. General formation of interfacial assembled hierarchical micro-nano arrays for biomass upgrading-coupled hydrogen production. Advanced Functional Materials. 2024;**34**(6):2308198. DOI: 10.1002/adfm.202308198

[35] Zhang D, Zhang Z, Meng L, Dong L, Shen Y, Wang Z, et al. Synergistic engineering of Ag doping and P vacancies in nanoflower-like Ni_2P for acrylamide-assisted electrolysis of seawater splitting. Chemical Engineering Journal. 2024;**497**:154465. DOI: 10.1016/j.cej.2024.154465

[36] Liu X, Xu M, Wang L, Cui X, Nie W, Wang Z, et al. Efficient hydrogen production from seawater via Fe-EL-modified catalysts: A biomass-coupled approach for sustainable energy storage. ChemSusChem. 2025;**18**(12):e202500070. DOI: 10.1002/cssc.202500070

[37] Zhang L, Wang Z, Qiu J. Energy-saving hydrogen production by seawater electrolysis coupling sulfion degradation. Advanced Materials. 2022;**34**(16):2109321. DOI: 10.1002/adma.202109321

[38] Sun F, Qin J, Wang Z, Yu M, Wu X, Sun X, et al. Energy-saving hydrogen

production by chlorine-free hybrid seawater splitting coupling hydrazine degradation. Nature Communications. 2021;**12**(1):4182. DOI: 10.1038/s41467-021-24529-3

[39] Guo L, Chi J, Zhu J, Cui T, Lai J, Wang L. Dual-doping $NiMoO_4$ with multi-channel structure enable urea-assisted energy-saving H2 production at large current density in alkaline seawater. Applied Catalysis B: Environmental. 2023;**320**:121977. DOI: 10.1016/j.apcatb.2022.121977

[40] Chen H, Zhang K, Xia Y, Li J, Hui B. (B,P,Co,Fe)-Ni modified on nanowood for boosting seawater urea electro-oxidation. Green Chemistry. 2024;**27**(1):133-143. DOI: 10.1039/d4gc05156d